AF360912

Analysis and Design of Structures Made of Plastically Anisotropic Materials

Analysis and Design of Structures Made of Plastically Anisotropic Materials

Editors

Sergei Alexandrov
Lihui Lang

MDPI • Basel • Beijing • Wuhan • Barcelona • Belgrade • Manchester • Tokyo • Cluj • Tianjin

Editors
Sergei Alexandrov
Beihang University
China
Ishlinsky Institute for Problems
in Mechanics of the Russian
Academy of Sciences
Russia

Lihui Lang
Beihang University
China

Editorial Office
MDPI
St. Alban-Anlage 66
4052 Basel, Switzerland

This is a reprint of articles from the Special Issue published online in the open access journal *Symmetry* (ISSN 2073-8994) (available at: https://www.mdpi.com/journal/symmetry/special_issues/Analysis_Design_Structures_Made_Plastically_Anisotropic_Materials).

For citation purposes, cite each article independently as indicated on the article page online and as indicated below:

LastName, A.A.; LastName, B.B.; LastName, C.C. Article Title. *Journal Name* **Year**, *Article Number*, Page Range.

ISBN 978-3-03936-838-9 (Hbk)
ISBN 978-3-03936-839-6 (PDF)

Contents

About the Editors

Sergei Alexandrov is a research professor at the Institute for Problems of Mechanics of the Russian Academy of Science (Moscow, Russia) and a visiting professor at Beihang University (Beijing, China). He received his Ph.D. in Physics and Mathematics in 1990 and DSc in Physics and Mathematics in 1994. He worked as a professor at Moscow Aviation Technology Technical University (Russia), a visiting scientist at ALCOA Technical Center (USA), GKSS Research Centre (Germany), Seoul National University (South Korea) and National Chung Cheng University (Taiwan), and a visiting professor at National San Yat Sen University (Taiwan) University of Franche-Comte (France). His research areas are plasticity theory, fracture mechanics, and their applications to metal forming and structural mechanics. He is a member of the Russian National Committee on Theoretical and Applied Mechanics.

Lihui Lang obtained his Ph.D. degree from Harbin Institute of Technology in 1998. He is working now as a full professor and scientific committee member of Stamping and Forging Technology, Plasticity Engineering. His research interests are mainly focused on automotive, aircraft and aerospace fields, and his team research covers hydroforming including sheet hydroforming and tube hydroforming, fiber metal laminates composite, high temperature/pressure forming of powder and gradient function/structure materials, powder technologies, warm/hot forming/hydroforming of lightweight materials, high-efficiency exchanger, and KBE system. He has published more than 200 papers in journals, most of which are cited by SCI and EI. He has published one technical book, "Innovative Hydroforming and Warm/Hot Hydroforming". His research has been supported by the NICHIDAI Die Manufacturing Company Youth Prize. He was awarded the "Thomas Stephen Prize" by the Mechanical Engineering Society of the UK and has obtained more than 30 patents.

Preface to "Analysis and Design of Structures Made of Plastically Anisotropic Materials"

The present monograph contains seven chapters written by authors from several countries. All of these chapters are devoted anisotropic properties of materials. Considering anisotropic material properties is important because these properties significantly affect the behavior of structures and parameters of deformation processes. The corresponding theoretical and experimental methods should be capable of capturing this influence of material anisotropy. The present monograph summarizes new trends and established approaches in the mechanics of anisotropic materials.

Sergei Alexandrov, Lihui Lang

Editors

Article

Residual Stress Analysis of an Orthotropic Composite Cylinder under Thermal Loading and Unloading

Somayeh Bagherinejad Zarandi [1], Hsiang-Wei Lai [1], Yun-Che Wang [1,*] and Sergey Aizikovich [2,3]

[1] Department of Civil Engineering, National Cheng Kung University, Tainan 70101, Taiwan;
    somayehbz14@gmail.com (S.B.Z.); sgshou0418@gmail.com (H.-W.L.)
[2] Research and Education Center "Materials", Don State Technical University, Rostov-on-Don 344000, Russia;
    saizikovich@gmail.com
[3] Vorovich Research Institute of Mechanics and Applied Mathematics, Southern Federal University,
    Rostov-on-Don 344090, Russia
* Correspondence: yunche@mail.ncku.edu.tw

Received: 31 January 2019; Accepted: 27 February 2019; Published: 4 March 2019

Abstract: Elastoplastic analysis of a composite cylinder, consisting of an isotropic elastic inclusion surrounded by orthotropic matrix, is conducted via numerical parametric studies for examining its residual stress under thermal cycles. The matrix is assumed to be elastically and plastically orthotropic, and all of its material properties are temperature-dependent (TD). The Hill's anisotropic plasticity material model is adopted. The interface between the inclusion and matrix is perfectly bonded, and the outer boundary of the cylinder is fully constrained. A quasi-static, uniform temperature field is applied to the cylinder, which is analyzed under the plane-strain assumption. The mechanical responses of the composite cylinder are strongly affected by the material symmetry and temperature-dependent material properties. When the temperature-independent material properties are assumed, larger internal stresses at the loading phase are predicted. Furthermore, considering only yield stress being temperature dependent may be insufficient since other TD material parameters may also affect the stress distributions. In addition, plastic orthotropy inducing preferential yielding along certain directions leads to complex residual stress distributions when material properties are temperature-dependent.

Keywords: orthotropic plasticity; residual stress; temperature-dependent material properties; composite cylinder; finite element analysis

---

1. Introduction

Temperature effects on the plastic deformation have significant industrial and academic interests [1,2]. However, many studies in the literature make assumptions that the material properties are isotropic and temperature-independent (TI). For example, aluminum composite discs under thermal loading have been studied without using temperature-dependent (TD) material properties [3,4]. Considerations of elastic and plastic anisotropy are important when the deformation of textured metals or single crystals under thermal loading are in question [5]. Orthotropic plasticity material models have been extensively developed by Hill [6–8]. Anisotropic plasticity theory have been applied in many studies to understand directional dependent yielding phenomena. For example, Yoon10 et al. conducted research on the calibration of parameters used in anisotropic yield criterion from experimental tests in strongly textured aluminum sheets [9]. Numerical studies on predicting earing phenomena in anisotropic aluminum have been performed [10]. In addition, orthotropic plastic deformation in fiber-reinforced composite disc under spinning has been analyzed [11].

The importance of using temperature-dependent material properties in thermal loading analysis has been emphasized by Noda [12]. Thermomechanical responses of solid and hollow cylinders with the consideration of temperature-dependent material properties have been reported [13]. Realistic temperature functions to describe material properties at elevated temperatures for structural steels can be found in [14]. Elastoplastic stress analysis of thin discs with temperature-dependent material properties have been studied with analytical methods [15,16]. In addition, effects of thickness variations on the elastoplastic behavior of annular discs have been studied [17]. Although these studies consider the temperature-dependent material properties, they only deal with a system containing single material. Composite systems introduce additional complexity into the plasticity problem. Zarandi et al. examined the plastic responses of a composite disc, in two and three dimensions, under monotonic temperature loading with consideration of temperature dependent material properties [18]. In addition, the plasticity problem of a particular type of composite materials, termed functionally graded materials, under bending have recently been investigated [19]. The derived analytical solutions provide an efficient way in designing such materials.

In this work, we assume that the matrix material has orthogonal symmetry both in its elastic and plastic properties, such as a single crystal, metal with texture, or or fiber-reinforced composite materials. The matrix material is fully constrained on its outer rim, and its inner rim is in perfect bonding with an isotropic, purely elastic inclusion. The finite element method is adopted to conduct parametric studies on elastoplastic behavior and residual stress of the composite cylinder, analyzed under the plane-strain assumption, subjected to uniform, quasi-static thermal loading and unloading. Numerical schemes for solving elastoplasticity problems have been well established [20,21]. In this study, we conduct parametric studies to analyze residual stresses with software package [22]. Both temperature dependent and temperature independent material properties are considered. Effects of hardening and plane stress/strain are analyzed. Our numerical results may serve as reference data for experimental verifications or future analytical solutions to such a problem.

2. Theoretical and Numerical Considerations

As shown in Figure 1a, the composite cylinder consists of an isotropic, purely elastic inclusion and elastoplastic matrix with the orthotropic symmetry both in its elastic and plastic behavior. The inclusion-matrix interface is assumed to be perfectly bonded. The outer boundary of the composite cylinder is fully clamped, and a uniform temperature field is quasi-statically applied to the composite cylinder. Figure 1b shows representative thermal loading cycles. The thermal loading parameter serves as loading steps in our analysis. In the figure, ΔT at Points B, D and F is in ratio of 1:1.25:1.5. The temperature differences at the three points are 700, 875 and 1050 °C, respectively. During the thermal loading or unloading cycles, residual stress fields may be developed at Points C, E and G. The physical properties of the isotropic elastic inclusion (Young's modulus $E_i = 411$ GPa, Poisson's ratio $\nu_i = 0.28$, linear thermal expansion coefficient $\alpha_i = 5.0 \times 10^{-6}$ K^{-1}) are assumed to be temperature independent, but those of the elastoplastic matrix are temperature dependent. Physically it is envisioned that the inclusion is made of a ceramic material ($0 < r < a$), surrounded by metallic material ($a < r < b$), where r is the radial component of the polar coordinate system to describe the points in the domain. The elastoplastic problems are numerically analyzed via the finite element method in two dimensions.

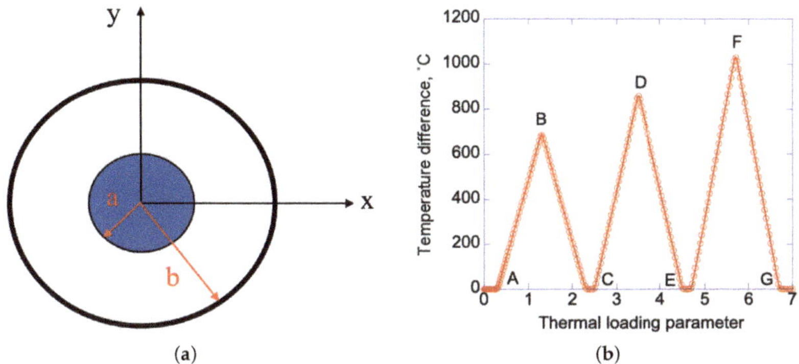

Figure 1. (**a**) Schematic of the composite cylinder, with $a = 0.3$ m and $b = 1$ m, confined on the outer rim, $r = 1$ m, and (**b**) representative quasi-static temperature loading profile.

In the elastic region, the orthotropic constitutive relationship for the matrix is as follows.

$$\varepsilon_{ij} = S_{ijkl}\sigma_{kl} \qquad \text{or} \qquad \varepsilon_m = S_{mn}\sigma_n \tag{1}$$

where, at the reference temperature, $E_{11} = 190$, $E_{22} = 200$, $E_{33} = 210$, $\nu_{12} = 0.25$, $\nu_{23} = 0.3$, $\nu_{13} = 0.35$, $G_{12} = 75$, $C_{12} - 80$, $G_{12} = 85$. Moduli are in units of GPa. Thermal expansion coefficient is assumed to have negligible orientational dependence, hence $\alpha_m = 11.7 \times 10^{-6}$ in units of $1/\text{K}$.

In the orthotropic symmetry, the Cartesian coordinates, (x_1, x_2, x_3) or (x, y, z), are the rolling direction, the transverse direction and the normal direction, respectively. The Hill's orthogonal yield function is defined as

$$\mathcal{F}(\sigma_{ij}) = F\left(\sigma_{22} - \sigma_{33}\right)^2 + G\left(\sigma_{33} - \sigma_{11}\right)^2 + H\left(\sigma_{11} - \sigma_{22}\right)^2 + 2L\sigma_{23}^2 + 2M\sigma_{13}^2 + 2N\sigma_{12}^2 \tag{2}$$

It is assumed that at the reference temperature yield stresses are $\sigma_{11} = 310$, $\sigma_{22} = 410$, $\sigma_{33} = 510$, $\sigma_{23} = 200$, $\sigma_{13} = 300$, $\sigma_{12} = 400$ in units of MPa. The associate flow rule is as follows.

$$\dot{\varepsilon}_{ij}^p = \lambda \frac{\partial \mathcal{F}}{\partial \sigma_{ij}} \tag{3}$$

Since the trance of the plastic strain tensor is zero due to the incompressibility assumption during plastic flow,

$$
\begin{aligned}
\dot{\varepsilon}_{11}^p &= 2\lambda\left[-G\left(\sigma_{33} - \sigma_{11}\right) + H\left(\sigma_{11} - \sigma_{22}\right)\right] && \text{(4)} \\
\dot{\varepsilon}_{22}^p &= 2\lambda\left[F\left(\sigma_{22} - \sigma_{33}\right) - H\left(\sigma_{11} - \sigma_{22}\right)\right] && \text{(5)} \\
\dot{\varepsilon}_{33}^p &= 2\lambda\left[-F\left(\sigma_{22} - \sigma_{33}\right) + G\left(\sigma_{33} - \sigma_{11}\right)\right] && \text{(6)}
\end{aligned}
$$

The consequence if the incompressibility assumption leads to,

$$\varepsilon_{11}^p + \varepsilon_{22}^p + \varepsilon_{33}^p = 0 \tag{7}$$

The inter-relationships among Hill's parameters and directional yield stresses are as follows.

$$\sigma_{y23} = \sqrt{\frac{1}{2L}}, \qquad \sigma_{y31} = \sqrt{\frac{1}{2M}}, \qquad \sigma_{y12} = \sqrt{\frac{1}{2N}} \tag{8}$$

and

$$F = \frac{1}{2}\left(\frac{1}{\sigma_{y22}^2} + \frac{1}{\sigma_{y33}^2} - \frac{1}{\sigma_{y11}^2}\right) \tag{9}$$

$$G = \frac{1}{2}\left(\frac{1}{\sigma_{y33}^2} + \frac{1}{\sigma_{y11}^2} - \frac{1}{\sigma_{y22}^2}\right) \tag{10}$$

$$H = \frac{1}{2}\left(\frac{1}{\sigma_{y11}^2} + \frac{1}{\sigma_{y22}^2} - \frac{1}{\sigma_{y33}^2}\right) \tag{11}$$

In this work, we specify the six yield stresses, instead the Hill's parameters. One may further define equivalent initial yield stress

$$\sigma_{y0} = \sqrt{\frac{3}{2(F+G+H)}} \tag{12}$$

hence the Hill's effective stress

$$\sigma_{Hill}^2 = \sigma_{y0}^2 \left[F\left(\sigma_{22} - \sigma_{33}\right)^2 + G\left(\sigma_{33} - \sigma_{11}\right)^2 + H\left(\sigma_{11} - \sigma_{22}\right)^2 + 2L\sigma_{23}^2 + 2M\sigma_{13}^2 + 2N\sigma_{12}^2\right] \tag{13}$$

The plastic potential used in the isotropic hardening

$$\mathcal{F}_h = \sigma_{Hill} - \sigma_y \tag{14}$$

where

$$\sigma_y = \sigma_{y0} + \sigma_h(\varepsilon_{ep}) \tag{15}$$

The hardening function σ_h depends on effective plastic strains ε_{ep}. In the present analysis,

$$\sigma_h(\varepsilon_{ep}) = E_{iso}\varepsilon_{ep} \tag{16}$$

where

$$\frac{1}{E_{iso}} = \frac{1}{E_{Tiso}} - \frac{1}{E} \tag{17}$$

and E_{Tiso} is the isotropic tangent modulus and E the effective Young's modulus if the material is elastically anisotropic. In this work, we set $E_{Tiso} = 20$ MPa if linear hardening is considered. The local effective plastic strain is defined as follows.

$$\dot{\varepsilon}_{ep} = \sqrt{\frac{2}{3}\dot{\varepsilon}_{ij}^p\dot{\varepsilon}_{ij}^p} \tag{18}$$

The von Mises effective stress is defined as follows in terms of deviatoric stress tensor s_{ij}, or its second invariant J_2 [5].

$$\sigma_{mises} = \sqrt{3J_2(s_{ij})} = \sqrt{\frac{3}{2}s_{ij}s_{ij}} \quad \text{and} \quad s_{ij} = \sigma_{ij} - \frac{1}{3}\sigma_{kk}\delta_{ij} \tag{19}$$

here δ_{ij} is the Kronecker delta function, and the Einstein summation rule for the indices is applied. The temperature functions, shown in Equations (20)–(23), are used in this study. They are similar to those in Argeso and Eraslan ([13]), but slightly modified.

$$f_\sigma(T) = \sigma_y(T)/\sigma_0 = 1 + T/[600 \times \ln(T/1630)] \tag{20}$$

$$f_E(T) = E(T)/E_0 = 1 + T/[2000 \times \ln(T/1800)] \tag{21}$$

$$f_\nu(T) = \nu(T)/\nu_0 = 1 + 2.5 \times 10^{-4}T - 2.5 \times 10^{-7}T^2 \tag{22}$$

$$f_\alpha(T) = \alpha(T)/\alpha_0 = 1 + 2.56 \times 10^{-4}T - 2.14 \times 10^{-7}T^2 \tag{23}$$

where the applied temperature difference T is in units of $^\circ$C and reference temperature is 22 $^\circ$C. A graphical representation of the temperature functions is shown in Figure 2. For the case that the material is elastically and plastically isotropic, at the reference temperature, the Young's modulus, yield stress, Poisson's ratio and linear thermal expansion coefficient of the matrix material are assumed to be $E_0 = 200$ GPa, $\sigma_0 = 410$ MPa, $\nu_0 = 0.3$, and $\alpha_0 = 11.7 \times 10^{-6}$ per $^\circ$C, respectively. When the material is orthotropic, the above-mentioned temperature functions are applied to the material parameters at the reference temperature, listed after Equation (1). In other words, temperature-dependent yield stress $\sigma(T) = \sigma_0 f_\sigma(T)$, Young's modulus $E(T) = E_0 f_E(T)$, Poisson's ratio $\nu(T) = \nu_0 f_\nu(T)$ and linear thermal expansion coefficient $\alpha(T) = \alpha_0 f_\alpha(T)$. This choice of material parameters is representative for studying the temperature dependent elastoplastic materials in general. The deformation process is assumed to be quasi-static throughout this work.

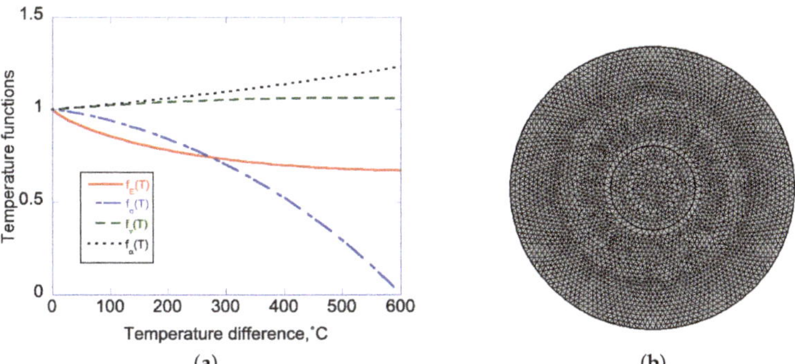

Figure 2. (**a**) Temperature functions for the material properties and (**b**) representative finite element mesh used in this study.

A representative finite element mesh is shown in Figure 2b with the inclusion radius $a = 0.3$ m and the radius to the outer boundary of the cylinder is $b = 1$ m. The number of two-dimensional quadratic serendipity elements used in the analysis was about 6000, and the number of degrees of freedom (d.o.f.) was about 170,000, including the internal d.o.f. for plasticity. We adopted COMSOL ([22]) software for the finite element calculations.

3. Results and Discussion

3.1. Residual Stress Analysis Under Elastic-Perfectly Plastic Assumption

When the plane-plane composite cylinder, with elastic-perfectly plastic model, under temperature cyclic loading, Figure 3 shows the residual stress at $r = 0.35$ m, near the inclusion-matrix boundary, in the TD case under the maximum applied temperature differences $\Delta T = 270, 337.5, 405\ ^\circ$C. All material parameters are assumed to be temperature dependent. The maximum applied temperature differences are labeled as the letters 'B', 'D' and 'F' in the Figure 1, as schematics. Since the chosen loading magnitudes are large, reversed plasticity occurs during unloading. The stress magnitudes at the polar angle $\theta = 45°$ direction are larger than those at $\theta = 0°$ due to the chosen plasticity parameters in the Hill's model having larger yield stress when $\theta = 45°$.

Under the TI assumption, Figure 4 shows the residual stress in the orthotropic cylinder under maximum temperature differences $\Delta T = 270, 337.5, 405 \ °C$, which are the same as those used in the TD case. Since the loading magnitudes are kept the same in Figures 3 and 4, direct comparisons to exhibit the the differences between TD and TI assumptions can be accomplished. In general, since TI does not reduce yield stress at high temperature, it predicts higher stresses. Furthermore, it can be seen that the residual stresses, at the loading parameters about 2.5, 4.5 and 7, are mildly developed during reversed plasticity in the TI case since at high temperatures yield stress is not deduced. These loading parameters respectively correspond to the letters 'C', 'E' and 'G', shown in the schematics in Figure 1b. However, in the TD case, yield stress is largely reduced at high temperatures as shown in Figure 2a. This large reduction in yield stress causes strong reversed plasticity in the unloading phases. Hence, the TD residual stresses in Figure 3, at the loading parameters about 2.5, 4.5 and 7, are much larger than those in the TI case.

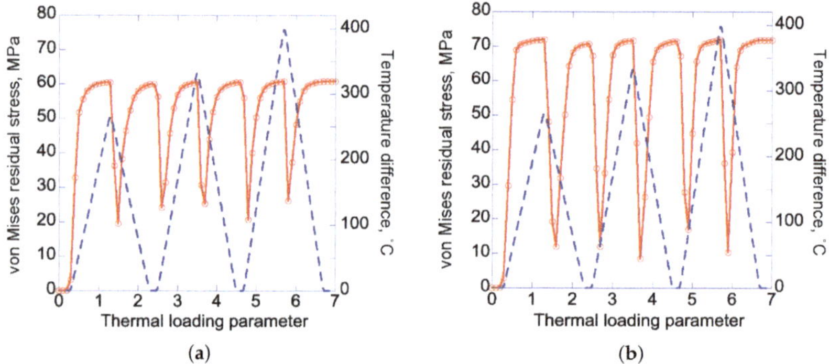

Figure 3. Von Mises residual stress at $r = 0.35$ m in the temperature-dependent (TD) case with the polar angle (**a**) $\theta = 0°$ and (**b**) $\theta = 45°$. Dashed lines indicate the applied temperature difference.

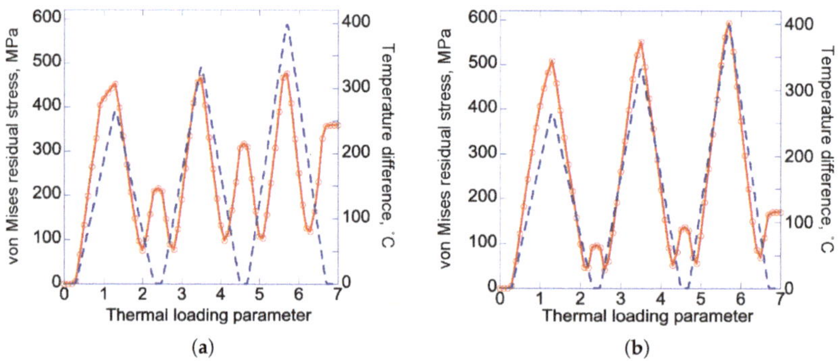

Figure 4. Von Mises residual stress at $r = 0.35$ m in the temperature-independent (TI) case with the polar angle (**a**) $\theta = 0°$ and (**b**) $\theta = 45°$. Dashed lines indicate the applied temperature difference.

3.2. Effects of Selective Temperature Dependent Material Properties on Residual Stress

In the previous section, the results are obtained with all material properties being temperature dependent. In this section, we examine the effects of selective TD material properties on residual stress. Linear hardening with $E_T = 20$ MPa is assumed and its temperature dependence is assumed to be

$f_E(T)$. Figure 5a shows the residual stress in the orthotropic cylinder after the maximum temperature difference $\Delta T = 400\ ^\circ\text{C}$ loading with all material parameters being temperature dependent, i.e., the four temperature functions listed in Equations (20)–(23) are used in the analysis. When only yield stress is considered to be temperature dependent, the corresponding residual stress is shown in Figure 5b. As can be seen their von Mises residual stress distributions are similar, but their magnitudes are distinct. Furthermore, the range for the residual stress 'plateau' is shorter when all material properties are temperature dependent. The 'plateau' is slightly inclined due to the small linear hardening E_T. As for comparisons, the residual stress distribution for the TI case is shown in Figure 5c. Due to no reduction in yield stress in the TI case as temperature increases, sharp residual stress distribution is developed near the inclusion-matrix interface. It is remarked that residual stresses are self-equilibrated inside the material. However, the von Mises residual stresses do not show this trend since they have averaged according to Equation (19).

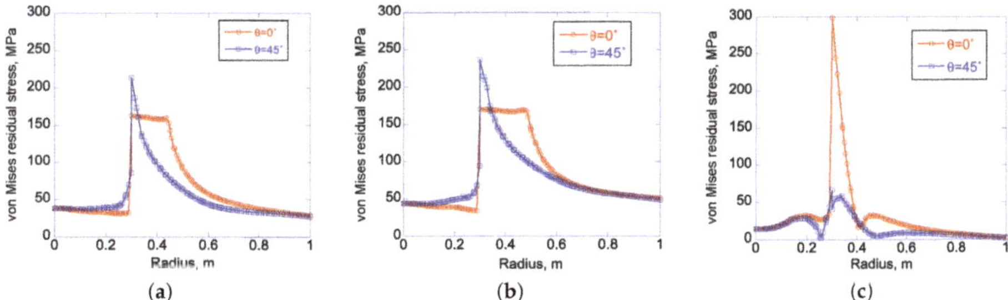

Figure 5. Residual stress distribution for (**a**) all material parameters being temperature dependent, (**b**) only yield stress being temperature dependent and (**c**) all material parameters being temperature independent.

3.3. Residual Stress Analysis with Linear Hardening

When the plane-strain composite cylinder under maximum temperature differences $\Delta T = 400$, 500, $600\ ^\circ\text{C}$, the residual stresses, at $r = 0.35$ m, in each loading cycle are shown in Figure 6. Both plastic anisotropy and isotropy are compared. Linear hardening is assumed, as in previous section. For the TD case, all material properties are assumed to be temperature dependent in this section. It can be seen that the TI case predicts large residual stress due to reversed plasticity in the unloading phase. Due to orthotropic elastic constants, the numerical values between the isotropic and orthotropic case can only be compared qualitatively.

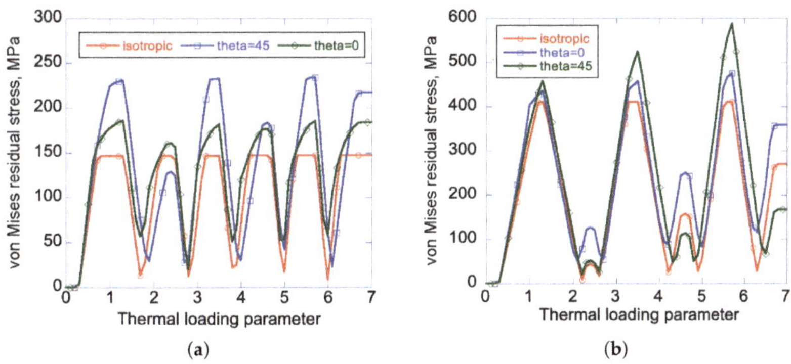

Figure 6. Residual stress near the inclusion-matrix interface at $r = 0.35$ for the (**a**) TD and (**b**) TI case.

To observe the orientational dependence of residual stress in the cylinder, Figure 7 shows the stress contour in the cylinder after unloading from maximum temperature difference $\Delta T = 400\,^\circ$C. Color bars indicate the von Mises residual stress in units of MPa. The TI case shows weaker developments in residual stress due to yield stress not being reduced at elevated temperature. Due to the matrix being both elastically and plastically orthotropic, the isotropic inclusion also shows Under the same loading/unloading conditions, Figure 8 shows the residual stress developed in the isotropic composite cylinder. As expected, no orientational dependence is observed in the residual stress distribution.

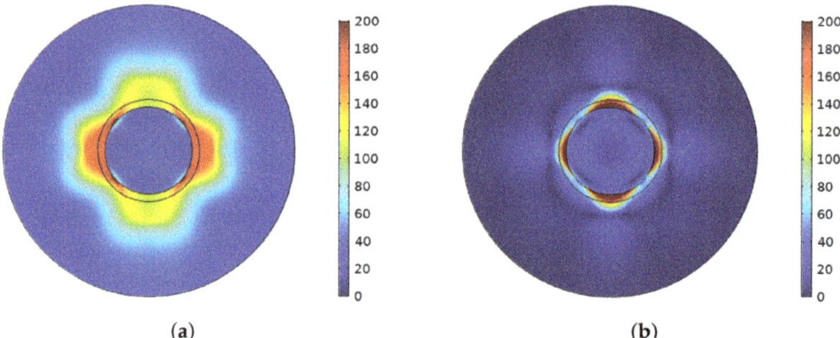

Figure 7. Von Mises residual stress in the orthotropic composite cylinder with the (**a**) TD and (**b**) TI material properties.

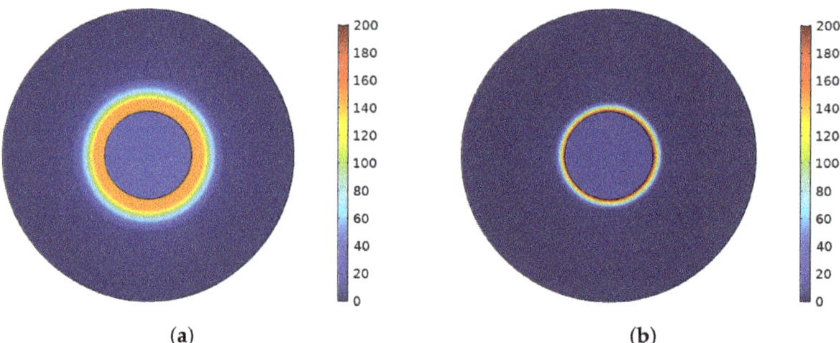

Figure 8. Von Mises residual stress in the isotropic composite cylinder with the (**a**) TD and (**b**) TI material properties.

In order to better observe the directional dependence of the residual stress, shown in Figure 7, we plot the residual stress at $r = 0.35$ m, near the inclusion-matrix interface indicated by a black circle in the figure, around the circumferential direction, as shown in Figure 9. The circumferential direction is indicated by the polar angle whose zero value is at the x-axis. It is remarked that for isotropic plasticity such a plot would show horizontal lines only; no orientational dependence. As can be seen in Figure 9, there are significant differences in the von Mises residual stress between the TD and TI case. In the TD case, due to significant reduction in yield stress, the reversed plasticity during unloading phase induces complex residual stress distribution.

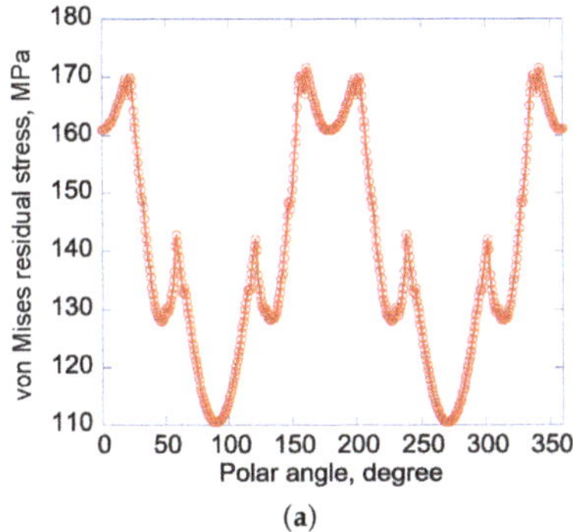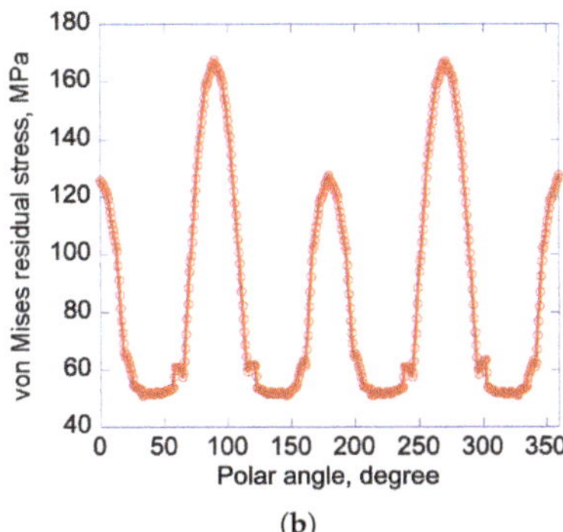

(a)   (b)

Figure 9. Von Mises residual stress at $r = 0.35$ m around the polar angle in the orthotropic cylinder for the (**a**) TD and (**b**) TI case.

Although many practical problems need to be analyzed in three space dimensions, long cylinders can be analyzed in two dimensions under the plane strain assumption. Through the basic research on the parametric analysis, it is demonstrated that the purpose of this study is to examine the residual stress in the orthotropic cylinder under thermal loading with the consideration of temperature dependent material properties. If materials have strong temperature dependence, one needs to be aware significant differences between the TD and TI case. In the future, considerations of temperature rates are required in order to realistically model the material responses.

Possible real-world applications of the present analysis are machine parts with anisotropic characteristics under repeated temperature loading/unloading cycles or textured alloys during metal forming processes. In addition, fiber-reinforced composite materials with fibers being arranged as concentric rings may require the consideration of both elastic and plastic orthotropy, as studied in the present work. This type of composite materials may be used in machinery, civil engineering or biomedical engineering.

The development of residual stresses in materials depend on a variety of variables [23]. If the residual stresses are to be minimized for certain applications, suitable selection of materials and boundary conditions, as well as the methodology of analyzing the residual stresses under given conditions, need to be considered. From the present study, to avoid complex residual stress developments in the TD case under thermal loading, one may consider to choose a material with less anisotropy and temperature sensitivity. For the TI case, one may either choose more isotropic material or use multiple layers near the interface to reduce the orientational dependence in residual stress. The multiple layers with suitable design and material selections may reduce the stress in the matrix, hence reduce yielding and consequent the developments of residual stresses after unloading. The multiple layer method may also work for the TD case. Further analysis is required.

4. Conclusions

Our parametric studies demonstrate that material symmetry plays an important role in the residual stress distribution in the composite cylinder under thermal excitation. The combination of material symmetry and temperature dependence in material properties may lead to complex residual stress patterns. In designing cylinders as components for engineering applications, plastic anisotropic effects cannot be ignored since they cause the orientational dependence in stress distribution along the circumferential coordinate. In addition, considerations of all material properties to be temperature dependent are important to reflect the physical processes in materials at high temperature. Residual stresses calculated from only considering yield stress to be temperature dependent are different from those with all TD material parameters. The assumptions of TI material properties do not reduce yield stress at high temperatures, hence larger internal stresses under loading are obtained when compared to the TD case. It is found that the TD case may give rise to complex residual stress distribution, as opposed to the TI case. Our numerical results and analysis procedure presented here may serve as a methodology generating numerical data for benchmark tests to compare with future analytical or experimental solutions.

Author Contributions: Conceptualization, S.B.Z., Y.-C.W. and S.A.; methodology, S.B.Z., H.-W.L. and Y.-C.W.; software, S.B.Z., H.-W.L. and Y.-C.W.; validation, S.B.Z., H.-W.L. and Y.-C.W.; formal analysis, Y.-C.W. and S.A.; investigation, S.B.Z., H.-W.L., Y.-C.W. and S.A.; resources, Y.-C.W. and S.A.; data curation, S.B.Z., H.-W.L. and Y.-C.W.; writing—original draft preparation, S.B.Z., H.-W.L., Y.-C.W. and S.A.; writing—review and editing, S.B.Z., H.-W.L., Y.-C.W. and S.A.; visualization, S.B.Z., H.-W.L. and Y.-C.W.; supervision, Y.-C.W. and S.A.; project administration, Y.-C.W. and S.A.; funding acquisition, Y.-C.W.

Funding: This research was funded, in part, by Ministry of Science and Technology, Taiwan, grant number MOST 107-2221-E-006-028.

Acknowledgments: This research was, in part, supported by the Ministry of Education, Taiwan, R.O.C, and the Aim for the Top University Project to the National Cheng Kung University (NCKU). Y.-C.W. is grateful to the National Center for High-performance Computing, Taiwan, for computer time and facilities.

Conflicts of Interest: The authors declare no conflict of interest.

References

1. Alexandrov, S.; Chikanova, N. Elastic-plastic stress-strain state of a plate with a pressed-in inclusion in thermal field. *Mech. Solids* **2000**, *35*, 125–132.
2. Alexandrov, S.; Alexandrova, N. Thermal effects on the development of plastic zones in thin axisymmetric plates. *J. Strain Anal. Eng. Des.* **2001**, *36*, 169–175. [CrossRef]
3. Sayman, O.; Arman, Y. Thermal stresses in a thermoplastic composite disc under a steady state temperature distribution. *J. Rein. Plas. Comp.* **2006**, *25*, 1709–1722. [CrossRef]
4. Topcu, M.; Altan, G.; Callioglu, H.; Altan, B.D. Thermal elastic-plastic analysis of an aluminium composite disc under linearly decreasing thermal loading. *Adv. Comp. Lett.* **2008**, *17*, 87–96. [CrossRef]
5. Lubliner, J. *Plasticity Theory*; Macmillan: New York, NY, USA, 1990.
6. Hill, R. A theory of the yielding and plastic flow of anisotropic metals. *Soc. Lond. Ser. A* **1948**, *193*, 281–297.
7. Hill, R. Constitutive modelling of orthotropic plasticity in sheet metals. *J. Mech. Phys. Solids* **1990**, *38*, 405–417. [CrossRef]
8. Hill, R. A user-friendly theory of orthotropic plasticity in sheet metals. *Int. J. Mech. Sci.* **1993**, *35*, 19–25. [CrossRef]
9. Yoon, J.-H.; Cazacu, O.; Yoon, J.W.; Dick, R.E. Earing predictions for strongly textured aluminum sheets. *Intl. J. Mech. Sci.* **2010**, *52*, 1563–1578. [CrossRef]
10. Zhang, S.; Leotoing, L.; Guines, D.; Thuillier, S.; Zang, S.-L. Calibration of anisotropic yield criterion with conventional tests or biaxial test. *Intl. J. Mech. Sci.* **2014**, *85*, 142–151. [CrossRef]
11. Callioglu, H.; Topcu, M.; Tarakcilar, A.R. Elastic–plastic stress analysis of an orthotropic rotating disc. *Intl. J. Mech. Sci.* **2006**, *48*, 985–990. [CrossRef]
12. Noda, N. Thermal stresses in materials with temperature-dependent properties. *Appl. Mech. Rev.* **1991**, *44*, 383–397. [CrossRef]
13. Argeso, H.; Eraslan, A.N. On the use of temperature-dependent physical properties in thermomechanical calculations for solid and hollow cylinders. *Int. J. Thermal Sci.* **2008**, *47*, 136–146. [CrossRef]
14. Seif, M.; Main, J.; Weigand, J.; Sadek, F.; Choe, L.; Xhang, C.; Gross, J.; Luecke, W.; McColskey, D. Temperature-Dependent Material Modeling for Structural Steels: Formulation and Application. *NIST Technical Note 1907*, **2016**. [CrossRef]
15. Alexandrov, S.; Wang, Y.C.; Aizikovich, S. Effect of temperature-dependent mechanical properties on plastic collapse of thin discs. *J. Mech. Eng. Sci. Part C* **2014**, *228*, 2483–2487. [CrossRef]
16. Alexandrov, S.; Wang, Y.C.; Jeng, Y.R. Elastic-plastic stresses and strains in thin discs with temperature-dependent properties subject to thermal loading. *J. Therm. Stresses* **2014**, *37*, 488–505. [CrossRef]
17. Wang, Y.C.; Alexandrov, S.; Jeng, Y.R. Effects of thickness variations on the thermal elastoplastic behavior of annular discs. *Struct. Eng. Mech.* **2013**, *47*, 839–856. [CrossRef]

18. Zarandi, S.B.; Wang, Y.C.; Novozhilova, O.V. Plastic behavior of circular discs with temperature-dependent properties containing an elastic inclusion. *Str. Eng. Mech.* **2016**, *58*, 731–743. [CrossRef]

19. Alexandrov, S.; Wang, Y.C.; Lang, L. A theory of elastic/plastic plane strain pure bending of FGM sheets at large strain. *Materials* **2019**, *12*, 456. [CrossRef] [PubMed]

20. Simo, J.C. Numerical analysis and simulation of plasticity. In *Handbook of Numerical Analysis VI*; Ciarlet, P.G., Lions, J.L., Eds.; Elsevier Science: Amsterdam, The Netherlands, 1998; pp. 183–499.

21. Simo, J.C.; Hughes, T.J.R. *Computational Inelasticity*; Springer: New York, NY, USA, 1998.

22. COMSOL Website (2019). Available online: www.comsol.com (accessed on 27 February 2019).

23. Young, W.B. *Residual Stress in Design, Process and Materials Selections*; ASM Intl.: Materials Park, OH, USA, 1989.

Article

A Theory of Autofrettage for Open-Ended, Polar Orthotropic Cylinders

Marina Rynkovskaya [1],*, Sergei Alexandrov [2,3] and Lihui Lang [2]

[1] Department of Civil Engineering, Peoples' Friendship University of Russia (RUDN University),
6 Miklukho-Maklaya St, Moscow 117198, Russia
[2] School of Mechanical Engineering and Automation, Beihang University, No. 37 Xueyuan Road,
Beijing 100191, China; sergei_alexandrov@spartak.ru (S.A.); lang@buaa.edu.cn (L.L.)
[3] Ishlinsky Institute for Problems in Mechanics, 101-1 Prospect Vernadskogo, Moscow 119526, Russia
* Correspondence: rynkovskaya_mi@pfur.ru

Received: 1 January 2019; Accepted: 19 February 2019; Published: 22 February 2019

Abstract: Autofrettage is a widely used process to enhance the fatigue life of holes. In the theoretical investigation presented in this article, a semi-analytic solution is derived for a polar, orthotropic, open-ended cylinder subjected to internal pressure, followed by unloading. Numerical techniques are only necessary to solve a linear differential equation and evaluate ordinary integrals. The generalized Hooke's law connects the elastic portion of strain and stress. The flow theory of plasticity is employed. Plastic yielding is controlled by the Tsai–Hill yield criterion and its associated flow rule. It is shown that using the strain rate compatibility equation facilitates the solution. The general solution takes into account that elastic and plastic properties can be anisotropic. An illustrative example demonstrates the effect of plastic anisotropy on the distribution of stresses and strains, including residual stresses and strain, for elastically isotropic materials.

Keywords: residual stress; residual strain; open-ended cylinder; autofrettage

---

1. Introduction

High-pressure vessels are often autofrettaged to improve their performance under service conditions. Numerous theories of the autofrettage process of hollow cylinders under different end conditions are available. The three main end conditions are usually adopted (plane strain, closed-end, and open-end conditions). The earliest attempt on a strict mathematical theory of the autofrettage process appears to have been in [1], where the plane strain condition has been considered assuming an elastic, perfectly plastic material model. This theory has been extended to closed-end tubes in [2]. A theory of the autofrettage process of tubes with free ends has been proposed in [3]. The Tresca yield criterion has been adopted, and the solution has been found by a finite difference method.

The elastic/perfectly plastic solutions mentioned above have been extended to other constitutive equations. In particular, solutions for open-ended cylinders of strain-hardening material have been derived in [4,5]. Both the Tresca and von Mises criteria, in conjunction with the corresponding associated flow rule, have been adopted in [4]. In the case of Ni-Cr-Mo cylinders, it has been shown that the effect of strain hardening is important in cylinders with radius ratios of 3 or greater. Hencky's deformation theory of plasticity, based on the von Mises yield criterion, has been employed in [5]. A solution for hollow cylinders under a constant axial strain condition has been provided in [6], using the deformation theory of plasticity and the von Mises yield criterion. The corresponding plane strain solution can be obtained as a special case. A nonlinear strain-hardening model for steel has been proposed in [7]. Then, this model has been used for studying the process of autofrettage in close-ended cylinders. A comprehensive overview of autofrettage theories for internally pressurized homogeneous tubes of perfectly plastic and strain-hardening materials has been provided in [8]. A plane strain solution based on a gradient theory of plasticity has been found in [9]. Hencky's deformation theory of plasticity and a unified yield criterion have been adopted.

The Bauschinger effect can significantly influence the distribution of residual stresses and strains in tubes subjected to internal pressure followed by unloading. Therefore, many theoretical solutions for the process of autofrettage are based on material models that incorporate the Bauschinger effect. A solution for a hardening law suitable for high-strength steel has been given in [10]. A distinguished feature of this hardening law is that the material is perfectly plastic at loading, but shows a strong Bauschinger effect within a certain range of the forward strain. The Tresca yield criterion and its associated flow rule have been used. An approximate method of finding analytic solutions for generic isotropic and kinematic strain hardening laws has been introduced in [11]. Another approximate method has been employed in [12], using the concept of the single effective material. Numerical methods have been developed in [13–15] for materials with nonlinear stress–strain behavior. An effect of varying elastic and plastic material properties along the radius on the distribution of residual stresses in autofrettaged cylinders has been evaluated in [16].

An efficient method of improving the performance of autofrettaged tubes is to use two- and multi-layer tubes [17]. Several theoretical solutions for such tubes are available in the literature (for example, [18–23]). The methods of analysis employed are similar to those used for homogeneous tubes.

In addition to the autofrettage treatment by internal pressure, thermal and rotational autofrettage treatments are widely used. Thermal autofrettage has been studied in [24–27], and rotational autofrettage in [28,29].

A comprehensive overview of theoretical and experimental research on the process of autofrettage has been recently provided in [30]. It is seen from this review that initially anisotropic materials were not considered. On the other hand, it is known from solutions to other problems in structural mechanics, for example in [31–33], that plastic anisotropy may have a significant effect on the solution. In particular, it is mentioned in [33] that even mild plastic anisotropy significantly affects the distribution of residual stresses, which is of special importance for the process of autofrettage. In the case of circular discs and cylinders, a common type of anisotropy is polar orthotropy. In particular, the effect of plastic anisotropy on stress and strain fields in rotating discs has been studied in [34–39], using different material models and boundary conditions. Various boundary value problems for orthotropic cylinders have been solved in [40–44]. All of these studies demonstrate that it is important to take into account plastic anisotropy in analysis and the design of structures. It is therefore reasonable to provide a theoretical analysis of the autofrettage process for polar orthotropic cylinders.

In the present paper, the open-ended cylinder is considered. It is assumed that the elastic strain and stress are connected by the generalized Hooke's law. Plastic yielding is controlled by the Tsai–Hill yield criterion. This criterion is often used in applications [45–49]. Therefore, the material is initially anisotropic. The flow theory of plasticity is employed. It is shown that using the strain rate compatibility equation facilitates the solution. In particular, a numerical technique is only necessary to solve a linear differential equation and evaluate ordinary integrals.

2. Statement of the Problem

Consider the expansion of a thick-walled hollow cylinder of inner radius a_0 and outer radius b_0 by a uniform internal pressure P_0, followed by unloading. The external pressure is zero. It is natural to solve this boundary value problem in a cylindrical coordinate system (r, θ, z) whose $z-$axis coincides with the axis of symmetry of the cylinder. It is assumed that the cylinder is sufficiently long to make the stresses and strains independent of the z-coordinate. The ends of the cylinder are not loaded. The inner pressure at the end of loading is high enough so that the annulus contained by the inner radius and some internal radius $r = r_c$ is plastic, while the outer annulus contained by the surface $r = r_c$ and the outer radius is elastic. The surface $r = r_c$ is the elastic/plastic boundary. Let σ_r, σ_θ, and σ_z be the stress components referred to the cylindrical coordinate system. These stresses are the principal stresses. Moreover, $\sigma_z = 0$ for the open-ended cylinder. The boundary conditions at loading are

$$\sigma_r = -P_0 \tag{1}$$

for $r = a_0$, and

$$\sigma_r = 0 \tag{2}$$

for $r = b_0$. Let P_m be the value of P_0 at the end of loading. Then, the boundary conditions at unloading are

$$\Delta\sigma_r = P_m \tag{3}$$

for $r = a_0$, and

$$\Delta\sigma_r = 0 \tag{4}$$

for $r = b_0$. Here $\Delta\sigma_r$ is the increment of the radial stress in course of unloading.

It is assumed that the cylinder is polar orthotropic. Then, the principal strain directions coincide with the principal stress directions. In particular, the generalized Hooke's law, in terms of the principal stress and strain components under plane stress conditions, is

$$\varepsilon_r^e = a_{rr}\sigma_r + a_{r\theta}\sigma_\theta, \quad \varepsilon_\theta^e = a_{r\theta}\sigma_r + a_{\theta\theta}\sigma_\theta, \quad \varepsilon_z^e = a_{rz}\sigma_r + a_{\theta z}\sigma_\theta. \tag{5}$$

Here ε_r^e, ε_θ^e, and ε_z^e are the elastic radial, circumferential, and axial strains, respectively. The coefficients a_{rr}, $a_{r\theta}$, a_{rz}, and $a_{\theta z}$ are the components of the compliance tensor. In terms of the principal stresses, the Tsai–Hill yield criterion reads

$$\sigma_\theta^2 - \sigma_r\sigma_\theta + \sigma_r^2\frac{X^2}{Y^2} = X^2 \tag{6}$$

where X and Y are the yield stresses in the circumferential and radial directions, respectively. The flow rule associated with the yield criterion (6) is

$$\frac{\partial\varepsilon_r^p}{\partial t} = \lambda_1\left(\frac{2X^2}{Y^2}\sigma_r - \sigma_\theta\right), \quad \frac{\partial\varepsilon_\theta^p}{\partial t} = \lambda_1(2\sigma_\theta - \sigma_r), \quad \frac{\partial\varepsilon_z^p}{\partial t} = \lambda_1\left[\left(1 - \frac{2X^2}{Y^2}\right)\sigma_r - \sigma_\theta\right] \tag{7}$$

where ε_r^p, ε_θ^p, and ε_z^p are the plastic radial, circumferential, and axial strains, respectively; t is the time; and λ_1 is a non-negative multiplier. Since the model under consideration is rate independent, the time derivatives in (7) can be replaced with derivatives with respect to any monotonically increasing or decreasing parameter q. Then, Equation (7) is replaced with

$$\zeta_r^p = \lambda\left(\frac{2X^2}{Y^2}\sigma_r - \sigma_\theta\right), \quad \zeta_\theta^p = \lambda(2\sigma_\theta - \sigma_r), \quad \zeta_z^p = \lambda\left[\left(1 - \frac{2X^2}{Y^2}\right)\sigma_r - \sigma_\theta\right] \tag{8}$$

where $\zeta_r^p = \partial\varepsilon_r^p/\partial q$, $\zeta_\theta^p = \partial\varepsilon_\theta^p/\partial q$, $\zeta_z^p = \partial\varepsilon_z^p/\partial q$, and λ is proportional to λ_1. The total strains are given by

$$\varepsilon_r = \varepsilon_r^e + \varepsilon_r^p, \quad \varepsilon_\theta = \varepsilon_\theta^e + \varepsilon_\theta^p, \quad \varepsilon_z = \varepsilon_z^e + \varepsilon_z^p. \tag{9}$$

The constitutive equations should be supplemented with the equilibrium equation of the form

$$\frac{\partial\sigma_r}{\partial r} + \frac{\sigma_r - \sigma_\theta}{r} = 0. \tag{10}$$

The solution is facilitated by using the equation of strain-rate compatibility. This equation is equivalent to

$$r\frac{\partial\zeta_\theta}{\partial r} + \zeta_\theta - \zeta_r = 0. \tag{11}$$

In what follows, the following dimensionless quantities will be used:

$$\rho = \frac{r}{b_0}, \quad a = \frac{a_0}{b_0}, \quad \rho_c = \frac{r_c}{b_0}, \quad p_0 = \frac{P_0}{X}, \quad p_m = \frac{P_m}{X}, \quad k = Xa_{rr}. \tag{12}$$

3. Purely Elastic Solution

The general purely elastic solution for stress can be written as

$$\frac{\sigma_r}{X} = C_1\rho^{\tau-1} + C_2\rho^{-\tau-1}, \quad \frac{\sigma_\theta}{X} = \tau\left(C_1\rho^{\tau-1} - C_2\rho^{-\tau-1}\right) \tag{13}$$

where C_1 and C_2 are constants of integration and $\tau = \sqrt{a_{rr}/a_{\theta\theta}}$. Substituting Equation (13) into Equation (5) supplies the solution for strain in the form

$$\begin{aligned}
\frac{\epsilon_r^e}{k} &= C_1\left(1 + \frac{Xa_{r\theta}\tau}{k}\right)\rho^{\tau-1} + C_2\left(1 - \frac{Xa_{r\theta}\tau}{k}\right)\rho^{-\tau-1}, \\
\frac{\epsilon_\theta^e}{k} &= C_1\left(\frac{a_{r\theta}}{a_{rr}} + \frac{a_{\theta\theta}\tau}{a_{rr}}\right)\rho^{\tau-1} + C_2\left(\frac{a_{r\theta}}{a_{rr}} - \frac{a_{\theta\theta}\tau}{a_{rr}}\right)\rho^{-\tau-1}, \\
\frac{\epsilon_z^e}{k} &= C_1\left(\frac{a_{rz}}{a_{rr}} + \frac{\tau a_{\theta z}}{a_{rr}}\right)\rho^{\tau-1} + C_2\left(\frac{a_{rz}}{a_{rr}} - \frac{\tau a_{\theta z}}{a_{rr}}\right)\rho^{-\tau-1}.
\end{aligned} \tag{14}$$

The solution for Equation (13) should satisfy the boundary conditions of Equations (1) and (2). Then, using Equation (12), the constants C_1 and C_2 are determined as

$$C_1 = -\frac{p_0}{a^{t-1} - a^{-t-1}}, \quad C_2 = \frac{p_0}{a^{t-1} - a^{-t-1}}. \tag{15}$$

Substituting Equation (15) into Equation (13) results in

$$\frac{\sigma_r}{X} = \frac{p_0}{(a^{\tau-1} - a^{-\tau-1})}\left(\rho^{-\tau-1} - \rho^{\tau-1}\right), \quad \frac{\sigma_\theta}{X} = -\frac{\tau p_0}{(a^{\tau-1} - a^{-\tau-1})}\left(\rho^{-\tau-1} + \rho^{\tau-1}\right). \tag{16}$$

It is assumed that plastic yielding initiates at the inner radius of the cylinder, $\rho = a$. This assumption should be verified for each set of constitutive parameters. The corresponding condition follows from Equations (6) and (16), in the form

$$\frac{p_0^2}{(a^{\tau-1} - a^{-\tau-1})^2}\left[\tau^2\left(\rho^{-\tau-1} + \rho^{\tau-1}\right)^2 + \frac{\tau}{\rho^2}\left(\rho^{-2\tau} - \rho^{2\tau}\right) + \left(\rho^{-\tau-1} - \rho^{\tau-1}\right)^2\frac{X^2}{Y^2}\right] \leq 1 \tag{17}$$

in the range $a \leq \rho \leq 1$. It follows from Equation (16) that

$$\frac{\sigma_r}{X} = -p_0, \quad \frac{\sigma_\theta}{X} = \frac{\tau p_0(1 + a^{2\tau})}{(1 - a^{2\tau})} \tag{18}$$

at $\rho = a$. Substituting Equation (18) into the yield criterion of Equation (6) and using Equation (12) yields

$$p_e = \left(1 - a^{2\tau}\right)\left[\tau^2\left(1 + a^{2\tau}\right)^2 + \tau\left(1 - a^{4\tau}\right) + \frac{X^2}{Y^2}\left(1 - a^{2\tau}\right)^2\right]^{-1/2}. \tag{19}$$

Here p_e is the value of p_0, at which point a plastic region starts to propagate from the inner radius of the cylinder. In what follows, it is assumed that $p_0 > p_e$.

4. Elastic/Plastic Stress Solution

There are two regions, $a \leq \rho \leq \rho_c$ and $\rho_c \leq \rho \leq 1$, at $p_0 > p_e$. The region $\rho_c \leq \rho \leq 1$ is elastic. The general solution for Equation (13) is valid in this region. However, the constants C_1 and C_2 are not given by (15). The stress solution in the region $a \leq \rho \leq \rho_c$ must satisfy the yield criterion of Equation

(6) and the equilibrium Equation (10). It is possible to verify by inspection that the yield criterion is satisfied by the following substitution:

$$\frac{\sigma_r}{X} = -\frac{2\sin\varphi}{Q}, \quad \frac{\sigma_\theta}{X} = -\frac{\sin\varphi}{Q} - \cos\varphi, \quad Q = \frac{X}{Y}\sqrt{4 - \frac{Y^2}{X^2}} \tag{20}$$

where φ is an auxiliary function of ρ. Substituting Equation (20) into (10) yields

$$2\cos\varphi\frac{\partial\varphi}{\partial\rho} + \frac{(\sin\varphi - Q\cos\varphi)}{\rho} = 0. \tag{21}$$

The stress solution in the region $a \leq \rho \leq \rho_c$ should satisfy the boundary condition of Equation (1). Using Equations (12) and (20), this condition transforms to

$$\varphi = \varphi_a \tag{22}$$

where $\rho = a$, where φ_a is determined from the equation

$$2\sin\varphi_a = Qp_0 \tag{23}$$

The unique solution of this equation is found using the condition that the circumferential stress at $\rho = a$ at the initiation of plastic yielding is determined from Equation (18), in which p_0 should be replaced with p_e, given in Equation (19). The solution of Equation (21) satisfying the boundary condition of Equation (22) is

$$\ln\frac{\rho}{a} = \frac{2Q(\varphi - \varphi_a)}{(1 + Q^2)} + \frac{2}{(1 + Q^2)}\ln\left(\frac{Q\cos\varphi_a - \sin\varphi_a}{Q\cos\varphi - \sin\varphi}\right). \tag{24}$$

Let φ_c be the value of φ at $\rho = \rho_c$. Then, it follows from Equation (24) that

$$\ln\frac{\rho_c}{a} = \frac{2Q(\varphi_c - \varphi_a)}{(1 + Q^2)} + \frac{2}{(1 + Q^2)}\ln\left(\frac{Q\cos\varphi_a - \sin\varphi_a}{Q\cos\varphi_c - \sin\varphi_c}\right) \tag{25}$$

The solution of Equation (13) should satisfy the boundary condition in Equation (2). Therefore, using Equation (12), it is possible to find that $C_1 + C_2 = 0$. Then, the stress solution in the elastic region $\rho_c \leq \rho \leq 1$ is

$$\frac{\sigma_r}{X} = C_1\left(\rho^{\tau-1} - \rho^{-\tau-1}\right), \quad \frac{\sigma_\theta}{X} = \tau C_1\left(\rho^{\tau-1} + \rho^{-\tau-1}\right). \tag{26}$$

The radial and circumferential stresses must be continuous across the elastic/plastic boundary. Then, it follows from Equations (20) and (26) that

$$-\frac{2\sin\varphi_c}{Q} = C_1\left(\rho_c^{\tau-1} - \rho_c^{-\tau-1}\right), \quad -\frac{\sin\varphi_c}{Q} - \cos\varphi_c = \tau C_1\left(\rho_c^{\tau-1} + \rho_c^{-\tau-1}\right). \tag{27}$$

Eliminating C_1 between these equations results in

$$1 + Q\cot\varphi_c = \frac{2\tau\left(\rho_c^{2\tau} + 1\right)}{\left(\rho_c^{2\tau} - 1\right)}. \tag{28}$$

In this equation, ρ_c can be eliminated by means of Equation (25). The resulting equation can be solved numerically to find φ_c as a function of φ_a. Using this solution, ρ_c as a function of φ_a is immediate from Equation (25), and then C_1 is a function of φ_a from any part of Equations (27). Equation (23) allows for all these quantities to be expressed as a function of p_0. Then, at any value of p_0, the variation of stresses with ρ in the elastic region follows from Equation (26), and in the plastic region from (20) and (24). The latter is in parametric form, with φ being the parameter. A difficulty is that this general solution may not exist. One of the restrictions is that plastic yielding is not initiated in the elastic region. Using Equations (12), (6), and (26), the corresponding condition can be represented as

$$C_1^2\left[\tau^2\left(\rho^{\tau-1}+\rho^{-\tau-1}\right)^2 - \frac{\tau}{\rho^2}\left(\rho^{2\tau}-\rho^{-2\tau}\right) + \left(\rho^{\tau-1}-\rho^{-\tau-1}\right)^2\frac{X^2}{Y^2}\right] \le 1 \tag{29}$$

in the range $\rho_c \le \rho \le 1$. Having found the value of C_1 the inequality in Equation (29), it can be verified by inspection with no difficulty. Another restriction is immediate from (20):

$$\frac{Y}{X} < 2. \tag{30}$$

The physical sense of this restriction is that Equation (6) does not determine a convex yield surface in principal stress space if $Y > 2X$. Still another restriction follows from Equation (23). Since $|\sin \varphi_a| \le 1$, the value of p_0 must satisfy the inequality

$$p_0 \le \frac{2}{Q} \equiv p_p. \tag{31}$$

If $p_0 = p_p$, then the localization of plastic deformation occurs at the inner radius of the cylinder, and the plastic region cannot propagate beyond the radius reached at this value of p_0.

Consider the state of stress in the cylinder when the entire cylinder becomes plastic, and the localization of plastic deformation occurs at the inner radius of the cylinder simultaneously. The latter condition requires $\varphi_a = \pi/2$. On the other hand, the stresses in Equation (20) should satisfy the boundary condition in Equation (2). It is reasonable to assume that at $\sigma_\theta > 0$ at $\rho = 1$. Then, Equations (2) and (20) combine to give $\varphi_c = \pi$. It is evident that $\rho_c = 1$. Substituting $\varphi_a = \pi/2$, $\varphi_c = \pi$, and $\rho_c = 1$ into Equation (25) yields

$$\ln a = \frac{2\ln Q}{(1+Q^2)} - \frac{Q\pi}{(1+Q^2)}. \tag{32}$$

Here, Q can be eliminated using its definition. Then, Equation (32) determines a relationship between a and Y/X corresponding to the state of stress in question. This relation is illustrated in Figure 1. If the point corresponding to a pair $(a, Y/X)$ lies above the curve, then the entire disc becomes plastic before the localization of plastic deformation at the inner surface of the cylinder, and vice versa.

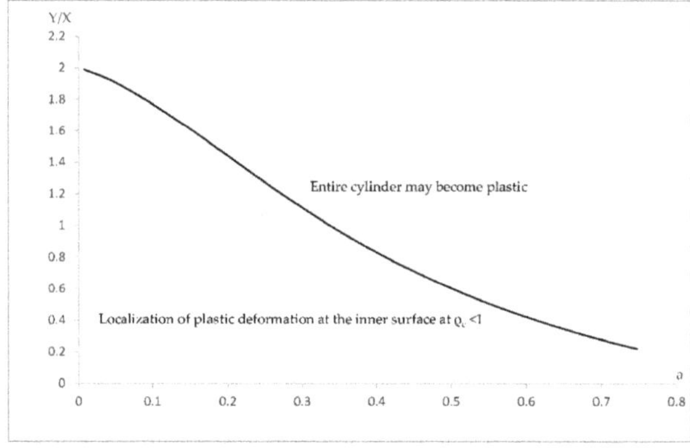

Figure 1. Geometric interpretation of two different mechanisms of plastic collapse (localization of plastic deformation at the inner radius of the cylinder and occurrence of the plastic region over the entire disc).

It is also of importance to consider the difference between p_e and p_p. It is seen from Equations (19) and (31) that p_p is the function only of Y/X, whereas p_e depends on Y/X, a, and τ. The variation of $p_p - p_e$ with Y/X at $a = 0.4$ for several values of τ is depicted in Figure 2. It is seen from this figure that the difference is rather small if the ratio Y/X is small enough. This means that the localization of plastic deformation at the inner surface of the cylinder occurs at the very beginning of plastic yielding.

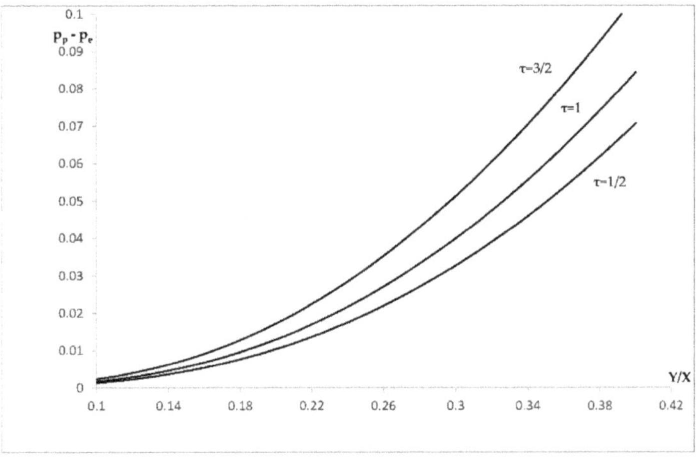

Figure 2. Effect of constitutive parameters on the magnitude of pressure at which plastic deformation is localized at the inner radius of the cylinder.

5. Elastic/Plastic Strain Solution

The total strain is elastic in the region $\rho_c \leq \rho \leq 1$. Therefore, using Equation (12), the principal strains in this region are found from the generalized Hooke's law in Equation (5) and the stress solution of Equation (26), as

$$
\begin{aligned}
\frac{\varepsilon_r}{k} &= C_1\left[\left(1 + \frac{\tau a_{r\theta}}{a_{rr}}\right)\rho^{\tau-1} + \left(\frac{\tau a_{r\theta}}{a_{rr}} - 1\right)\rho^{-\tau-1}\right], \\
\frac{\varepsilon_\theta}{k} &= C_1\left[\left(\frac{a_{r\theta}}{a_{rr}} + \frac{\tau a_{\theta\theta}}{a_{rr}}\right)\rho^{\tau-1} + \left(\frac{\tau a_{\theta\theta}}{a_{rr}} - \frac{a_{r\theta}}{a_{rr}}\right)\rho^{-\tau-1}\right], \\
\frac{\varepsilon_z}{k} &= C_1\left[\left(\frac{a_{rz}}{a_{rr}} + \frac{\tau a_{\theta z}}{a_{rr}}\right)\rho^{\tau-1} + \left(\frac{\tau a_{\theta z}}{a_{rr}} - \frac{a_{rz}}{a_{rr}}\right)\rho^{-\tau-1}\right].
\end{aligned}
\tag{33}
$$

Using Equation (12), the elastic portion of strain in the plastic region, $a \leq \rho \leq \rho_c$, is determined from the generalized Hooke's law (Equation (5)) and the stress solution in Equation (20), as

$$
\begin{aligned}
\frac{\varepsilon_r^e}{k} &= -\frac{(2 + a_{r\theta}/a_{rr})}{Q}\sin\varphi - \frac{a_{r\theta}}{a_{rr}}\cos\varphi, \qquad
\frac{\varepsilon_\theta^e}{k} = -\frac{(2 a_{r\theta} + a_{\theta\theta})}{Q a_{rr}}\sin\varphi - \frac{a_{\theta\theta}}{a_{rr}}\cos\varphi, \\
\frac{\varepsilon_z^e}{k} &= -\frac{2 a_{rz}\sin\varphi}{a_{rr} Q} - \frac{a_{\theta z}}{a_{rr}}\left(\frac{\sin\varphi}{Q} + \cos\varphi\right).
\end{aligned}
\tag{34}
$$

Substituting Equation (20) into Equation (8) leads to

$$
\begin{aligned}
\xi_r^p &= \lambda\left[\frac{\sin\varphi}{Q}\left(1 - \frac{4X^2}{Y^2}\right) + \cos\varphi\right], \\
\xi_\theta^p &= -2\lambda\cos\varphi, \qquad \xi_z^p = \lambda\left[\left(\frac{4X^2}{Y^2} - 1\right)\frac{\sin\varphi}{Q} + \cos\varphi\right].
\end{aligned}
\tag{35}
$$

Eliminating λ between these equations gives

$$
\begin{aligned}
\frac{\xi_r^p}{\xi_\theta^p} &= \frac{Q}{2}\left(\frac{4X^2}{Y^2} - 1\right)\tan\varphi - \frac{1}{2}, \\
\frac{\xi_z^p}{\xi_\theta^p} &= -\frac{Q}{2}\left(\frac{4X^2}{Y^2} - 1\right)\tan\varphi - \frac{1}{2}.
\end{aligned}
\tag{36}
$$

In what follows, it is assumed that $q \equiv \varphi_a$ and

$$\zeta_r^e = \frac{\partial \varepsilon_r^e}{\partial \varphi_a}, \quad \zeta_\theta^e = \frac{\partial \varepsilon_\theta^e}{\partial \varphi_a}, \quad \zeta_z^e = \frac{\partial \varepsilon_z^e}{\partial \varphi_a}, \quad \zeta_r = \frac{\partial \varepsilon_r}{\partial \varphi_a}, \quad \zeta_\theta^e = \frac{\partial \varepsilon_\theta}{\partial \varphi_a}, \quad \zeta_z^e = \frac{\partial \varepsilon_z}{\partial \varphi_a}. \tag{37}$$

Then, differentiating Equation (34) with respect to φ_a yields

$$\begin{aligned}
\frac{\zeta_r^e}{k} &= \left[-\frac{(2+a_{r\theta}/a_{rr})}{Q} \cos\varphi + \frac{a_{r\theta}}{a_{rr}} \sin\varphi \right] \frac{\partial\varphi}{\partial\varphi_a}, \\
\frac{\zeta_\theta^e}{k} &= \left[-\frac{(2a_{r\theta}+a_{\theta\theta})}{Qa_{rr}} \cos\varphi + \frac{a_{\theta\theta}}{a_{rr}} \sin\varphi \right] \frac{\partial\varphi}{\partial\varphi_a}, \\
\frac{\zeta_z^e}{k} &= -\left[\frac{2a_{rz}\cos\varphi}{a_{rr}Q} + \frac{a_{\theta z}}{a_{rr}} \left(\frac{\cos\varphi}{Q} - \sin\varphi \right) \right] \frac{\partial\varphi}{\partial\varphi_a}.
\end{aligned} \tag{38}$$

Substituting Equation (9) differentiated with respect to φ_a into Equation (11) and using Equation (12) leads to

$$\rho \frac{\partial \zeta_\theta^e}{\partial \rho} + \zeta_\theta - \zeta_r^p - \zeta_r^e = 0. \tag{39}$$

Moreover, using Equation (36),

$$\zeta_r^p = \left[\frac{Q}{2}\left(\frac{4X^2}{Y^2} - 1 \right) \tan\varphi - \frac{1}{2} \right] \zeta_\theta^p = \left[\frac{Q}{2}\left(\frac{4X^2}{Y^2} - 1 \right) \tan\varphi - \frac{1}{2} \right] (\zeta_\theta - \zeta_\theta^e) \tag{40}$$

Then, eliminating ζ_r^p in Equation (39) by means of Equation (40) yields

$$\rho \frac{\partial \zeta_\theta^e}{\partial \rho} + \frac{\zeta_\theta}{2}\left[3 - Q\left(\frac{4X^2}{Y^2} - 1 \right) \tan\varphi \right] + \left[\frac{Q}{2}\left(\frac{4X^2}{Y^2} - 1 \right) \tan\varphi - \frac{1}{2} \right] \zeta_\theta^e - \zeta_r^e = 0. \tag{41}$$

Using Equation (21), differentiation with respect to ρ in Equation (41) can be replaced with differentiation with respect to φ. As a result,

$$\frac{\partial \zeta_\theta^e}{\partial \varphi}(Q - \tan\varphi) + \zeta_\theta\left[3 - Q\left(\frac{4X^2}{Y^2} - 1 \right) \tan\varphi \right] + \left[Q\left(\frac{4X^2}{Y^2} - 1 \right) \tan\varphi - 1 \right] \zeta_\theta^e - 2\zeta_r^e = 0. \tag{42}$$

It is seen from (38) that the expressions for ζ_r^e and ζ_θ^e involve the derivative $\partial\varphi/\partial\varphi_a$. In general, this derivative can be found from Equation (24), which is the solution of Equation (21). However, it is more convenient to represent the solution of this equation satisfying the boundary condition Equation (22) as

$$\ln\frac{\rho}{a} = 2\int_{\varphi_a}^{\varphi} \frac{\cos\eta}{(Q\cos\eta - \sin\eta)} d\eta \tag{43}$$

where η is a dummy variable of integration. Differentiating Equation (43) gives

$$\frac{2\cos\varphi}{(Q\cos\varphi - \sin\varphi)} d\varphi = \frac{2\cos\varphi_a}{(Q\cos\varphi_a - \sin\varphi_a)} d\varphi_a + \frac{d\rho}{\rho}.$$

It follows from this equation that

$$\frac{\partial\varphi}{\partial\varphi_a} = \frac{\cos\varphi_a(Q\cos\varphi - \sin\varphi)}{\cos\varphi(Q\cos\varphi_a - \sin\varphi_a)}. \tag{44}$$

Equations (38) and (44) combine to give

$$
\begin{aligned}
\frac{\zeta^e_r}{k} &= \left[-\frac{(2+a_{r\theta}/a_{rr})}{Q}\cos\varphi + \frac{a_{r\theta}}{a_{rr}}\sin\varphi \right] \frac{(Q-\tan\varphi)}{(Q-\tan\varphi_a)}, \\
\frac{\zeta^e_\theta}{k} &= \left[-\frac{(2a_{r\theta}+a_{\theta\theta})}{Qa_{rr}}\cos\varphi + \frac{a_{\theta\theta}}{a_{rr}}\sin\varphi \right] \frac{(Q-\tan\varphi)}{(Q-\tan\varphi_a)}, \\
\frac{\zeta^e_z}{k} &= -\left[\frac{2a_{rz}\cos\varphi}{a_{rr}Q} + \frac{a_{\theta z}}{a_{rr}}\left(\frac{\cos\varphi}{Q} - \sin\varphi \right) \right] \frac{(Q-\tan\varphi)}{(Q-\tan\varphi_a)}.
\end{aligned}
\tag{45}
$$

Eliminating ζ^e_r and ζ^e_θ in Equation (42) by means of Equation (45) results in the following linear differential equation for ζ_θ/k:

$$
\begin{aligned}
&\frac{\partial(\zeta_\theta/k)}{\partial\varphi} + \frac{\zeta_\theta}{k}\Phi_1(\varphi) + \frac{\Phi_2(\varphi)}{(Q-\tan\varphi_a)} = 0, \\
&\Phi_1(\varphi) = \left[3 - Q\left(\frac{4X^2}{Y^2} - 1 \right)\tan\varphi \right](Q-\tan\varphi)^{-1},
\end{aligned}
$$

$$
\begin{aligned}
\Phi_2(\varphi) = {}&\left[a_{\theta\theta}\sin\varphi - \frac{(2a_{r\theta}+a_{\theta\theta})}{Q}\cos\varphi \right]\left[Q\left(\frac{4X^2}{Y^2} - 1 \right)\tan\varphi - 1 \right] + \\
&2\left[\frac{(2a_{rr}+a_{r\theta})}{Q}\cos\varphi - a_{r\theta}\sin\varphi \right].
\end{aligned}
\tag{46}
$$

The circumferential strain rate must be continuous across the elastic/plastic boundary. Therefore, the boundary condition to Equation (46) is

$$
\frac{\zeta_\theta}{k} = \frac{\zeta_c}{k}
\tag{47}
$$

for $\varphi = \varphi_c$. Here, ζ_c is the value of ζ_θ on the elastic side of the elastic/plastic boundary. Differentiating the second equation in Equation (33) with respect to φ_a, and then putting $\rho = \rho_c$ results in

$$
\frac{\zeta_c}{k} = \frac{dC_1}{d\varphi_a}\left[\left(\frac{a_{r\theta}}{a_{rr}} + \frac{\tau a_{\theta\theta}}{a_{rr}} \right)\rho_c^{\tau-1} + \left(\frac{\tau a_{\theta\theta}}{a_{rr}} - \frac{a_{r\theta}}{a_{rr}} \right)\rho_c^{-\tau-1} \right].
\tag{48}
$$

It is seen from this equation that it is necessary to find the derivative $dC_1/d\varphi_a$. It follows from Equation (43) that

$$
\ln\frac{\rho_c}{a} = 2\int_{\varphi_a}^{\varphi_c} \frac{\cos\varphi}{(Q\cos\varphi - \sin\varphi)}\,d\varphi.
\tag{49}
$$

Differentiating this equation and Equation (28) with respect to φ_a yields

$$
\frac{d\varphi_c}{d\varphi_a} = \frac{(Q-\tan\varphi_c)}{2}\left[\frac{d\rho_c}{\rho_c d\varphi_a} + \frac{2}{(Q-\tan\varphi_a)} \right]
\tag{50}
$$

and

$$
\frac{d\varphi_c}{d\varphi_a} = \frac{8\tau^2\rho_c^{2\tau-1}\sin^2\varphi_c}{Q(\rho_c^{2\tau}-1)^2}\frac{d\rho_c}{d\varphi_a},
\tag{51}
$$

respectively. Solving Equations (50) and (51) for the derivatives $d\rho_c/d\varphi_a$ and $d\varphi_c/d\varphi_a$ gives

$$
\begin{aligned}
\frac{d\rho_c}{d\varphi_a} &= \frac{(Q-\tan\varphi_c)}{(Q-\tan\varphi_a)}\left[\frac{8\tau^2\rho_c^{2\tau-1}\sin^2\varphi_c}{Q(\rho_c^{2\tau}-1)^2} - \frac{(Q-\tan\varphi_c)}{2\rho_c} \right]^{-1}, \\
\frac{d\varphi_c}{d\varphi_a} &= \frac{8\tau^2\rho_c^{2\tau-1}\sin^2\varphi_c(Q-\tan\varphi_c)}{Q(\rho_c^{2\tau}-1)^2(Q-\tan\varphi_a)}\left[\frac{8\tau^2\rho_c^{2\tau-1}\sin^2\varphi_c}{Q(\rho_c^{2\tau}-1)^2} - \frac{(Q-\tan\varphi_c)}{2\rho_c} \right]^{-1}.
\end{aligned}
\tag{52}
$$

The derivative $dC_1/d\varphi_a$ is determined from the first equation in Equation (27) as

$$
\frac{dC_1}{d\varphi_a} = \frac{2\sin\varphi_c\left[(\tau-1)\rho_c^{\tau-2} + (\tau+1)\rho_c^{-\tau-2} \right]}{Q\left(\rho_c^{\tau-1} - \rho_c^{-\tau-1} \right)^2}\frac{d\rho_c}{d\varphi_a} - \frac{2\cos\varphi_c}{Q\left(\rho_c^{\tau-1} - \rho_c^{-\tau-1} \right)}\frac{d\varphi_c}{d\varphi_a}
\tag{53}
$$

In this equation, the derivatives $d\rho_c/d\varphi_a$ and $d\varphi_c/d\varphi_a$ can be eliminated by means of Equation (52). In the previous section, φ_c and ρ_c have been found as functions of φ_a. Therefore, Equations (48) and (53) combine to supply ζ_c/k as a function of φ_a. Then, the solution of Equation (46), satisfying the boundary condition of Equation (47), can be solved numerically.

By definition, $\zeta_\theta = \partial\varepsilon_\theta/\partial\varphi_a$ if ζ_θ and ε_θ are regarded as functions of φ_a and ρ. However, the solution of Equation (46) provides ζ_θ as a function of φ_a and φ. In this case, $\frac{\partial\varepsilon_\theta}{\partial\varphi_a} + \frac{\partial\varepsilon_\theta}{\partial\varphi}\frac{\partial\varphi}{\partial\varphi_a} = \zeta_\theta$.

In this equation, the derivative $\partial\varphi/\partial\varphi_a$ can be eliminated by means of Equation (44). Then,

$$\frac{\partial\varepsilon_\theta}{\partial\varphi_a} + \frac{\partial\varepsilon_\theta}{\partial\varphi}\frac{(Q - \tan\varphi)}{(Q - \tan\varphi_a)} = \zeta_\theta. \tag{54}$$

Using a standard technique, it is possible to find that the equation of the characteristics is

$$d\varphi = \frac{(Q - \tan\varphi)}{(Q - \tan\varphi_a)}d\varphi_a \tag{55}$$

and the relation along the characteristics is

$$d\left(\frac{\varepsilon_\theta}{k}\right) = \frac{\zeta_\theta}{k}d\varphi_a. \tag{56}$$

Equation (55) can be immediately integrated to give

$$Q(\varphi_a - \varphi) + \ln\left(\frac{Q\cos\varphi - \sin\varphi}{Q\cos\varphi_a - \sin\varphi_a}\right) = D \tag{57}$$

where D is a constant of integration. The boundary condition to Equation (56) is that $\varepsilon_\theta/k = \varepsilon_\theta^e/k$ at the elastic/plastic boundary. Here ε_θ^e is the circumferential strain on the elastic side of the elastic/plastic boundary. Using Equation (33), this boundary condition is represented as

$$\frac{\varepsilon_\theta}{k} = C_1\left[\left(\frac{a_{r\theta}}{a_{rr}} + \frac{\tau a_{\theta\theta}}{a_{rr}}\right)\rho_c^{\tau-1} + \left(\frac{\tau a_{\theta\theta}}{a_{rr}} - \frac{a_{r\theta}}{a_{rr}}\right)\rho_c^{-\tau-1}\right] \tag{58}$$

for $\rho = \rho_c$ (or $\varphi = \varphi_c$).

It is evident from Equation (57) that $\varphi = \varphi_a$ is a characteristic curve, and that $D = 0$ on this curve. Having ζ_θ/k as a function of φ_a at $\varphi = \varphi_a$ (or $\rho = a$) from the solution of Equation (46), it is possible to integrate Equation (56) along the characteristic curve $\varphi = \varphi_a$ with the use of the boundary condition in Equation (58), to find the circumferential strain at the inner radius of the cylinder without solving Equation (54) for the entire plastic region. In order to illustrate the procedure for finding the strain solution in the entire plastic region, consider a schematic field of characteristics shown in Figure 3, where φ_m is the value of φ_a at the end of loading. Since φ_c as a function of φ_a is found from the solution of Equation (28), the curve $\varphi = \varphi_c$ is known. Choosing any pair (φ_a, φ) on this curve, it is possible to find D from Equation (57). The corresponding characteristic curve follows from Equation (57) at this value of D if φ_a varies in the range $\varphi_e \geq \varphi_a \geq \varphi_m$. In particular, the value of φ at $\varphi_a = \varphi_m$ is determined. This value of φ is denoted as φ_M. The value of the circumferential strain at $\varphi_a = \varphi_M$ and $\varphi = \varphi_M$ is found from the solution of Equation (56) satisfying the boundary condition of Equation (58). The plastic portion of this strain is immediate from Equations (9) and (34). Having found the distribution of ζ_θ/k along the characteristic curve, it is possible to determine the distribution of ζ_θ^p/k using the equation $\zeta_\theta^p/k = \zeta_\theta/k - \zeta_\theta^e/k$ and Equation (45). Then, Equation (36) supplies the distribution of ζ_r^p/k and ζ_z^p/k.

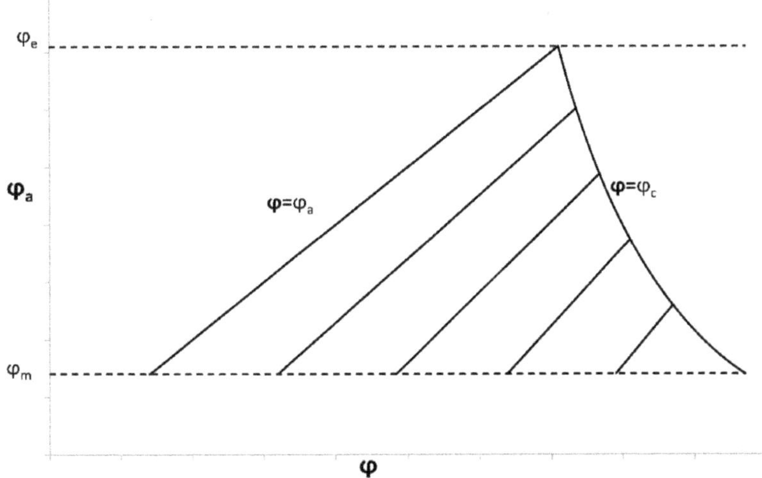

Figure 3. A schematic diagram showing the field of characteristics.

By analogy to Equation (54), it is possible to get

$$\frac{\partial \varepsilon_r^p}{\partial \varphi_a} + \frac{\partial \varepsilon_z^p}{\partial \varphi}\frac{(Q - \tan \psi)}{(Q - \tan \varphi_a)} = \zeta_\theta^p, \quad \frac{\partial c_z^p}{\partial \varphi_a} + \frac{\partial c_z^p}{\partial \varphi}\frac{(Q - \tan \varphi)}{(Q - \tan \varphi_a)} = \zeta_z^p. \tag{59}$$

These equations can be integrated in the same manner as Equation (54). In particular, Equation (57) is the equation of characteristic curves. The boundary conditions are

$$\varepsilon_r^p = \varepsilon_z^p = 0 \tag{60}$$

for $\rho = \rho_c$ (or $\varphi = \varphi_c$). Once the values of ε_r^p and ε_z^p at $\varphi_a = \varphi_m$ and $\varphi = \varphi_M$ have been found, the total strains are immediate from Equations (9) and (34). The strain solution described supplies the variation of strain components with φ at a given value of φ_a. In order to find the radial distributions, it is necessary to use Equation (24).

6. Unloading

It is assumed that the process of unloading is purely elastic. This assumption should be verified *a posteriori*. The general elastic solution of Equation (13), in which the stress components are replaced with their increments, is valid in the entire cylinder. Then,

$$\frac{\Delta\sigma_r}{X} = C_3\rho^{\tau-1} + C_4\rho^{-\tau-1}, \quad \frac{\Delta\sigma_\theta}{X} = \tau\left(C_3\rho^{\tau-1} - C_4\rho^{-\tau-1}\right) \tag{61}$$

where C_3 and C_4 are new constants of integration. These constants are found from the boundary conditions of Equations (3) and (4). As a result,

$$C_3 = -C_4 = \frac{p_m}{\left(a^{-\tau-1} - a^{\tau-1}\right)}. \tag{62}$$

Here, Equation (12) has been taken into account. Substituting Equation (62) into (61) supplies the radial distribution of $\Delta\sigma_r$ and $\Delta\sigma_\theta$ in the form

$$\frac{\Delta\sigma_r}{X} = \frac{p_m}{\left(a^{-\tau-1} - a^{\tau-1}\right)}\left(\rho^{\tau-1} - \rho^{-\tau-1}\right), \quad \frac{\Delta\sigma_\theta}{X} = \frac{\tau p_m}{\left(a^{-\tau-1} - a^{\tau-1}\right)}\left(\rho^{\tau-1} + \rho^{-\tau-1}\right). \tag{63}$$

The variation of the residual stresses with ρ is found as

$$\sigma_r^{res} = \sigma_r + \Delta\sigma_r \text{ and } \sigma_\theta^{res} = \sigma_\theta + \Delta\sigma_\theta. \tag{64}$$

It is understood here that σ_r and σ_θ are known from the stress solution given in Section 4, at $p_0 = p_m$. The process of unloading is purely elastic if the yield criterion is not violated in the entire cylinder. Using Equation (6), this condition can be represented as

$$\left(\frac{\sigma_\theta^{res}}{X}\right)^2 - \left(\frac{\sigma_\theta^{res}}{X}\right)\left(\frac{\sigma_r^{res}}{X}\right) + \left(\frac{\sigma_r^{res}}{X}\right)^2 \frac{X^2}{Y^2} \leq 1 \tag{65}$$

in the range $a \leq \rho \leq 1$. The radial distribution of the strain increments is determined from the generalized Hooke's law in Equations (5) and (62), as

$$
\begin{aligned}
\frac{\Delta\varepsilon_r^e}{k} &= \frac{p_m}{\left(a^{-\tau-1} - a^{\tau-1}\right)}\left[\left(1 + \frac{\tau a_{r\theta}}{a_{rr}}\right)\rho^{\tau-1} - \rho^{-\tau-1}\left(1 - \frac{\tau a_{r\theta}}{a_{rr}}\right)\right], \\
\frac{\Delta\varepsilon_\theta^e}{k} &= \frac{p_m}{\left(a^{-\tau-1} - a^{\tau-1}\right)}\left[\left(\frac{a_{r\theta}}{a_{rr}} + \frac{\tau a_{\theta\theta}}{a_{rr}}\right)\rho^{\tau-1} - \left(\frac{a_{r\theta}}{a_{rr}} - \frac{\tau a_{\theta\theta}}{a_{rr}}\right)\rho^{-\tau-1}\right], \\
\frac{\Delta\varepsilon_z^e}{k} &= \frac{p_m}{\left(a^{-\tau-1} - a^{\tau-1}\right)}\left[\left(\frac{a_{rz}}{a_{rr}} + \frac{\tau a_{\theta z}}{a_{rr}}\right)\rho^{\tau-1} - \left(\frac{a_{rz}}{a_{rr}} - \frac{\tau a_{\theta z}}{a_{rr}}\right)\rho^{-\tau-1}\right].
\end{aligned}
\tag{66}
$$

The variation of the residual strains with ρ is found as

$$\varepsilon_r^{res} = \varepsilon_r + \Delta\varepsilon_r, \ \varepsilon_\theta^{res} = \varepsilon_\theta + \Delta\varepsilon_\theta \text{ and } \varepsilon_z^{res} = \varepsilon_z + \Delta\varepsilon_z \tag{67}$$

It is understood here that ε_r, ε_θ, and ε_z are known from the strain solution given in Section 5 at $p_0 = p_m$.

7. Numerical Example

This section illustrates the effect of plastic anisotropy on the distribution of stress and strain in an $a = 0.4$ cylinder, assuming that the elastic properties are isotropic. In particular, it is assumed that Poisson's ratio is equal to 0.3 (i.e., $a_{r\theta} = -0.3$). The value of k is immaterial, because all strains are proportional to k. The solution given in Section 4 has been used to calculate the radial distribution of the radial and circumferential stress corresponding to $\rho_c = 0.8$. It is seen from Figure 1 that the solution without the localization of plastic deformation at the inner radius of the cylinder exists only if $Y/X > 0.8$. Therefore, the stress solution has been found at $Y/X = 0.85$, $Y/X = 1$ (isotropic material), $Y/X = 1.25$, and $Y/X = 1.5$. This solution is illustrated in Figure 4 (radial stress) and Figure 5 (circumferential stress). The associate strain solution has been found using the approach described in Section 5. This strain solution is illustrated in Figure 6 (total radial strain), Figure 7 (total circumferential strain), and Figure 8 (total axial strain). It can be seen from these figures that the effect of the ratio Y/X on the distribution of the strains is very significant in the range $Y/X < 1.25$. In this range, the magnitude of strains is very large in the vicinity of the inner surface of the cylinder, which indicates the tendency towards the localization of plastic deformation. Since the solution found is for small strains, it is necessary to verify for each combination of material and geometric parameters that the assumption of small strain is acceptable. The distribution of the residual stresses has been determined using the stress distributions depicted in Figures 4 and 5, in conjunction with the solution provided in Section 6. This solution is illustrated in Figure 9 (residual radial stress) and Figure 10 (residual circumferential stress). The associate strain solution has been found using the approach described in Section 6. This solution for residual strains is illustrated in Figure 11 (residual radial strain), Figure 12 (residual circumferential strain), and Figure 13 (residual axial strain). As in the case of the strain distribution at the end of loading, it is seen from these figures that the solution is very sensitive to the value of Y/X in the range $Y/X < 1.25$. The residual circumferential stress is of special significance for autofrettage. It is seen from Figure 10 that the magnitude of this stress at the inner surface of the cylinder is significantly affected by plastic anisotropy.

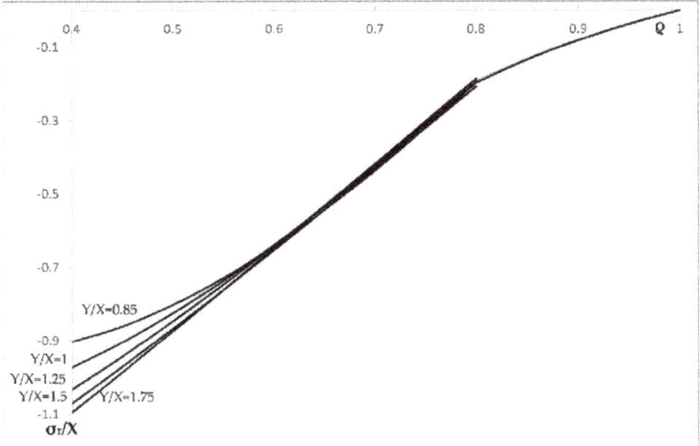

Figure 4. Variation of the radial stress with ρ in an a = 0.4 cylinder at several values of Y/X.

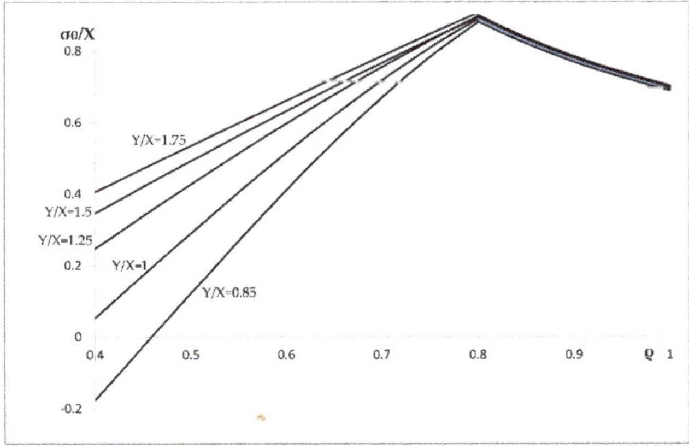

Figure 5. Variation of the circumferential stress with ρ in an a = 0.4 cylinder at several values of Y/X.

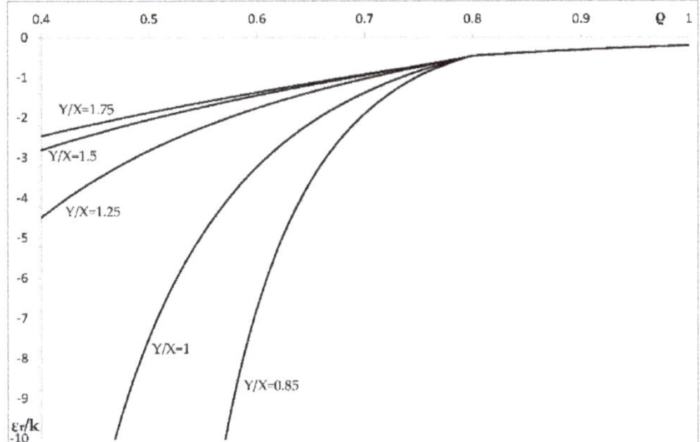

Figure 6. Variation of the total radial strain with ρ in an a = 0.4 cylinder at several values of Y/X.

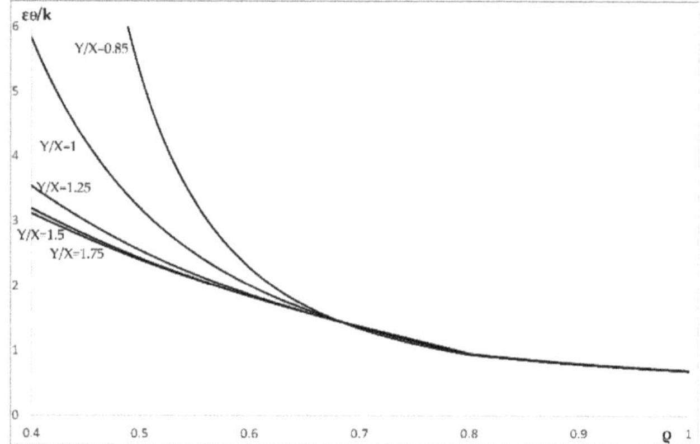

Figure 7. Variation of the total circumferential strain with ρ in an a = 0.4 cylinder at several values of Y/X.

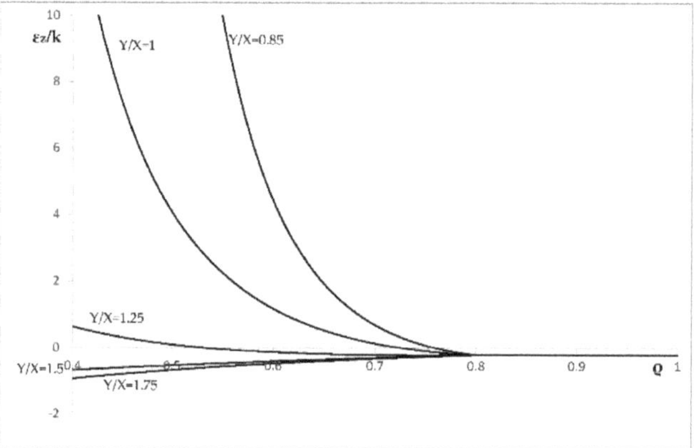

Figure 8. Variation of the total axial strain with ρ in an a = 0.4 cylinder at several values of Y/X.

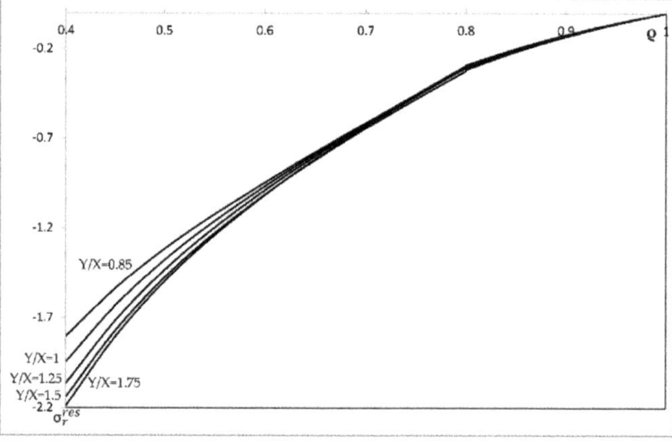

Figure 9. Variation of the residual radial stress with ρ in an a = 0.4 cylinder at several values of Y/X.

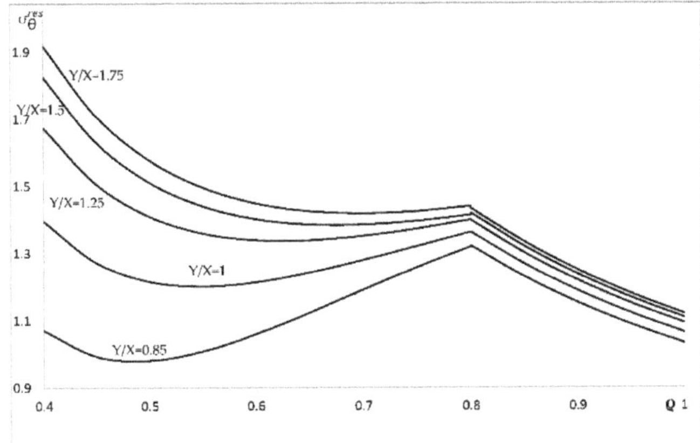

Figure 10. Variation of the residual circumferential stress with ρ in an a = 0.4 cylinder at several values of Y/X.

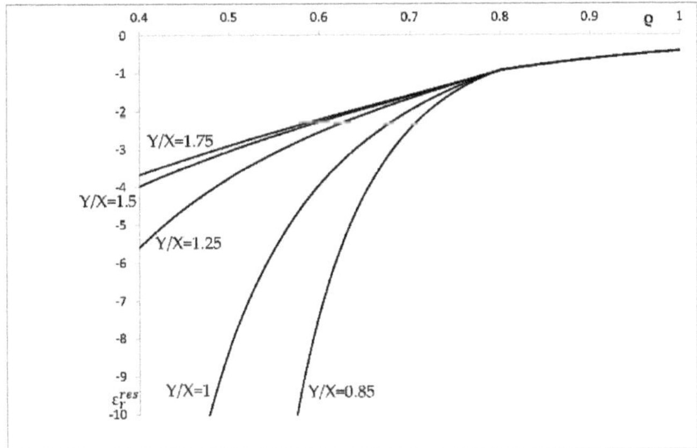

Figure 11. Variation of the residual radial strain with ρ in an a = 0.4 cylinder at several values of Y/X.

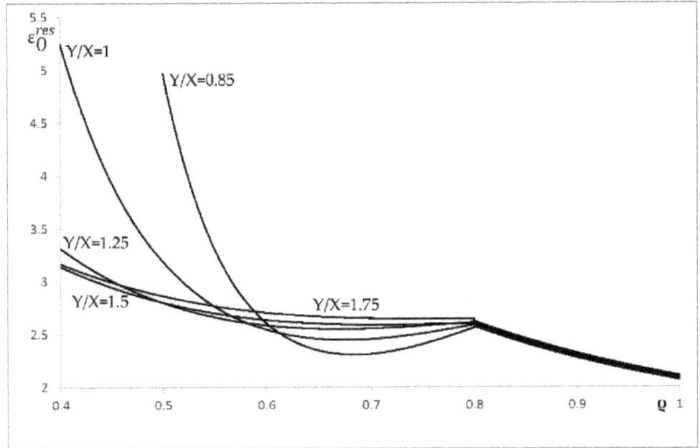

Figure 12. Variation of the residual circumferential strain with ρ in an a = 0.4 cylinder at several values of Y/X.

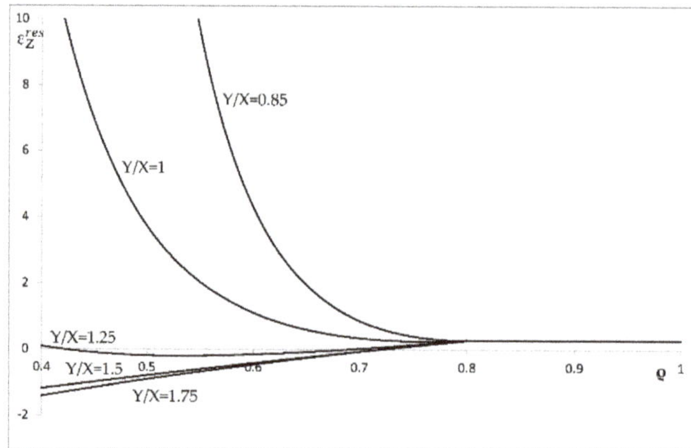

Figure 13. Variation of the residual axial strain with ρ in an a = 0.4 cylinder at several values of Y/X.

8. Conclusions

A new theoretical solution for the distribution of residual stresses and strains in an open-ended, thick-walled cylinder subjected to internal pressure followed by unloading has been proposed. A distinguished feature of this solution is that the cylinder is initially anisotropic. In particular, the paper is concentrated on a common type of anisotropy: polar orthotropy of elastic and plastic properties. The elastic response of the cylinder is controlled by the generalized Hooke's law, and the plastic response by the Tsai–Hill yield criterion and its associated flow rule. The flow theory of plasticity is employed. It has been shown that using the strain rate compatibility equation facilitates the solution. In particular, numerical techniques are only necessary to solve the linear differential Equation (46), and to evaluate ordinary integrals along characteristic curves.

The solution found can be directly used for the analysis and design of the process of autofrettage. It is worthy of note that in this case, there is no need to construct the field of strain in the entire cylinder, which is the most difficult part of the numerical solution. It follows from Equation (57) that $\varphi = \varphi_a$ is a characteristic curve, and this curve corresponds to the inner surface of the cylinder. The circumferential strain along this curve can be immediately found from Equation (56). Therefore, the radius of the cylinder after unloading is determined. The circumferential stress at the inner radius of the cylinder at the end of loading follows from Equation (20) at $\varphi = \varphi_a$. Then, the corresponding residual stress is immediate from Equations (61), (62), and (64).

An illustrative example is given in Section 7. In this case, it is assumed that the elastic properties are isotropic. As a result, the effect of the ratio Y/X on the distribution of stresses and strains has been revealed. This effect is especially significant in the range $Y/X < 1.25$ (Figures 5–8 and Figures 10–13). An exception is the distribution of the radial stress at the end of loading and after unloading. (Figures 4 and 9). This is because the boundary conditions on σ_r and $\Delta\sigma_r$, from Equations (2) and (94), dictate that this stress vanishes at the inner radius of the cylinder.

Author Contributions: All three authors participated in the research and in the writing of this paper.

Funding: S.A. acknowledges support from the Russian Foundation for Basic Research (Project 16-08-00469).

Acknowledgments: This work was initiated while M.R. was a visiting researcher at Beihang University, Beijing, China. The publication has been prepared with the support of the "RUDN University Program 5-100".

Conflicts of Interest: The authors declare no conflict of interest.

References

1. Hill, R.; Lee, E.H.; Tupper, S.J. The theory of combined plastic and elastic deformation with particular reference to a thick tube under internal pressure. *Proc. Roy. Soc. London. Series A Math. Phys. Sci.* **1947**, *191*, 278–303.
2. Hill, R. *The Mathematical Theory of Plasticity*; Clarendon Press: Oxford, UK, 1950.
3. Thomas, D.G.B. The autofrettage of thick tubes with free ends. *J. Mech. Phys. Solids* **1953**, *1*, 124–133. [CrossRef]
4. Rees, D.W.A. Autofrettage theory and fatigue life of open-ended cylinders. *J. Strain Anal.* **1990**, *25*, 109–121. [CrossRef]
5. Gao, X. An exact elasto-plastic solution for an open-ended thick-walled cylinder of a strain-hardening material. *Int. J. Press. Vessels Pip.* **1992**, *52*, 129–144. [CrossRef]
6. Hosseinian, E.; Farrahi, G.H.; Movahhedy, M.R. An analytical framework for the solution of autofrettaged tubes under constant axial strain condition. *J. Press. Vessel Techn.* **2009**, *131*, 1–8. [CrossRef]
7. Molaie, M.; Darijani, H.; Bahreman, M.; Hosseini, S.M. Autofrettage of nonlinear strain-hardening cylinders using the proposed analytical solution for stresses. *Int. J. Mech. Sci.* **2018**, *141*, 450–460. [CrossRef]
8. Rees, D.W.A. A theory for swaging of discs and lugs. *Meccanica* **2011**, *46*, 1213–1237. [CrossRef]
9. Gao, X.-L.; Wen, J.-F.; Xuan, F.-Z.; Tu, S.-T. Autofrettage and shakedown analyses of an internally pressurized thick-walled cylinder based on strain gradient plasticity solutions. *J. Appl. Mech.* **2015**, *82*, 1–12. [CrossRef]
10. Chen, P.C.T. The Bauschinger and Hardening effect on residual stresses in an autofrettaged thick-walled cylinder. *Press. Vessel Techn.* **1986**, *108*, 108–112. [CrossRef]
11. Livieri, P.; Lazzarin, P. Autofrettaged cylindrical vessels and bauschinger effect: An analytical frame for evaluating residual stress distributions. *J. Press. Vessel Techn.* **2002**, *124*, 38–46. [CrossRef]
12. Parker, A.P.; Gibson, M.C.; Hameed, A.; Troiano, E. Material modeling for autofrettage stress analysis including the "single effective material". *J. Press. Vessel Techn.* **2012**, *134*, 1–7. [CrossRef]
13. Gibson, M.C.; Parker, A.P.; Hameed, A.; Hetherington, J.G. Implementing realistic, nonlinear, material stress–strain behavior in ANSYS for the autofrettage of thick-walled cylinders. *J. Press. Vessel Techn.* **2012**, *134*, 1–7. [CrossRef]
14. Perl, M.; Perry, J. The beneficial influence of bauschinger effect mitigation on the barrel's safe maximum pressure. *J. Press. Vessel Techn.* **2013**, *135*, 1–5. [CrossRef]
15. Farrahi, G.H.; Voyiadjis, G.Z.; Hoseini, S.H.; Hosseinian, E. Residual stress analysis of the autofrettaged thick-walled tube using nonlinear kinematic hardening. *J. Press. Vessel Techn.* **2013**, *135*, 1–8. [CrossRef]
16. Haghpanah Jahromi, B.; Farrahi, G.H.; Maleki, M.; Nayeb-Hashemia, H.; Vaziri, A. Residual stresses in autofrettaged vessel made of functionally graded material. *Eng. Struct.* **2009**, *31*, 2930–2935. [CrossRef]
17. Jahed, H.; Farshi, B.; Karimi, M. Optimum autofrettage and shrink-fit combination in multi-layer cylinders. *J. Press. Vessel Techn.* **2006**, *128*, 196–200. [CrossRef]
18. Lee, E.-Y.; Lee, Y.-S.; Yang, Q.-M.; Kim, J.-H.; Cha, K.-U.; Hong, S.-K. Autofrettage process analysis of a compound cylinder based on the elastic-perfectly plastic and strain hardening stress-strain curve. *J. Mech. Sci. Techn.* **2009**, *23*, 3153–3160. [CrossRef]
19. Gexia, Y.; Hongzhao, L. An analytical solution of residual stresses for shrink-fit two-layer cylinders after autofrettage based on actual material behavior. *J. Press. Vessel Techn.* **2012**, *134*, 1–8. [CrossRef]
20. Benghalia, G.; Wood, J. Material and residual stress considerations associated with the autofrettage of weld clad components. *Int. J. Press. Vessels Pip.* **2016**, *139–140*, 146–158. [CrossRef]
21. Abdelsalam, O.R.; Sedaghati, R. Design optimization of compound cylinders subjected to autofrettage and shrink-fitting processes. *J. Press. Vessel Techn.* **2013**, *135*, 1–11. [CrossRef]
22. Hu, C.; Yang, F.; Zhao, Z.; Zeng, F. An alternative design method for the double-layer combined die using autofrettage theory. *Mech. Sci.* **2017**, *8*, 267–276. [CrossRef]
23. Seifi, R. Maximizing working pressure of autofrettaged three layer compound cylinders with considering Bauschinger effect and reverse yielding. *Meccanica* **2018**, *53*, 2485–2501. [CrossRef]
24. Hamilton, N.R.; Wood, J.; Easton, D.; Olsson Robbie, M.B.; Zhang, Y.; Galloway, A. Thermal autofrettage of dissimilar material brazed joints. *Mater. Des.* **2015**, *67*, 405–412. [CrossRef]
25. Kamal, S.M.; Dixit, U.S. Feasibility study of thermal autofrettage of thick-walled cylinders. *J. Press. Vessel Techn.* **2015**, *137*, 1–18. [CrossRef]

26. Kamal, S.M.; Borsaikia, A.C.; Dixit, U.S. Experimental assessment of residual stresses induced by the thermal autofrettage of thick-walled cylinders. *J. Strain Anal.* **2016**, *51*, 144–160. [CrossRef]

27. Shufen, R.; Dixit, U.S. An analysis of thermal autofrettage process with heat treatment. *Int. J. Mech. Sci.* **2018**, *144*, 134–145. [CrossRef]

28. Zare, H.R.; Darijani, H. A novel autofrettage method for strengthening and design of thick-walled cylinders. *Mater. Des.* **2016**, *105*, 366–374. [CrossRef]

29. Kamal, S.M. Analysis of residual stress in the rotational autofrettage of thick-walled disks. *J. Press. Vessel Techn.* **2018**, *140*, 1–10. [CrossRef]

30. Shufen, R.; Dixit, U.S. A review of theoretical and experimental research on various autofrettage processes. *ASME J. Press. Vessel Technol.* **2018**, *140*, 050802. [CrossRef]

31. Alexandrov, S.; Chung, K.-H.; Chung, K. Effect of plastic anisotropy of weld on limit load of undermatched middle cracked tension specimens. *Fat. Fract. Engng. Mater. Struct* **2007**, *30*, 333–341. [CrossRef]

32. Alexandrov, S.; Mustafa, Y. Influence of plastic anisotropy on the limit load of highly under-matched scarf joints with a crack subject to tension. *Eng. Fract. Mech.* **2014**, *131*, 616–626. [CrossRef]

33. Prime, M.B. Amplified effect of mild plastic anisotropy on residual stress and strain anisotropy. *Int. J. Solids Struct.* **2017**, *118*, 70–77. [CrossRef]

34. Alexandrova, N.; Alexandrov, S. Elastic-plastic stress distribution in a plastically anisotropic rotating disk. *Trans. ASME J. Appl. Mech.* **2004**, *71*, 427–429. [CrossRef]

35. Alexandrova, N.; Vila Real, P.M.M. Effect of plastic anisotropy on stress-strain field in thin rotating disks. *Thin-Walled Struct.* **2006**, *44*, 897–903. [CrossRef]

36. Peng, X.-L.; Li, X.-F. Elastic analysis of rotating functionally graded polar orthotropic disks. *Int. J. Mech. Sci.* **2012**, *60*, 84–91. [CrossRef]

37. Essa, S.; Argeso, H. Elastic analysis of variable profile and polar orthotropic FGM rotating disks for a variation function with three parameters. *Acta Mech.* **2017**, *228*, 3877–3899. [CrossRef]

38. Jeong, W.; Alexandrov, S.; Lang, L. Effect of plastic anisotropy on the distribution of residual stresses and strains in rotating annular disks. *Symmetry* **2018**, *10*, 420. [CrossRef]

39. Yildirim, V. Numerical/analytical solutions to the elastic response of arbitrarily functionally graded polar orthotropic rotating discs. *J. Brazilian Soc. Mech. Sci. Eng.* **2018**, *40*, 320. [CrossRef]

40. Leu, S.-Y.; Hsu, H.-C. Exact solutions for plastic responses of orthotropic strain-hardening rotating hollow cylinders. *Int. J. Mech. Sci.* **2010**, *52*, 1579–1587. [CrossRef]

41. Abd-Alla, A.M.; Mahmoud, S.R.; AL-Shehri, N.A. Effect of the rotation on a non-homogeneous infinite cylinder of orthotropic material. *Appl. Math. Comp.* **2011**, *217*, 8914–8922. [CrossRef]

42. Lubarda, V.A. On Pressurized curvilinearly orthotropic circular disk, cylinder and sphere made of radially nonuniform material. *J. Elast.* **2012**, *109*, 103–133. [CrossRef]

43. Croccolo, D.; De Agostinis, M. Analytical solution of stress and strain distributions in press fitted orthotropic cylinders. *Int. J. Mech. Sci.* **2013**, *71*, 21–29. [CrossRef]

44. Shahani, A.R.; Torki, H.S. Determination of the thermal stress wave propagation in orthotropic hollow cylinder based on classical theory of thermoelasticity. *Cont. Mech. Thermodyn.* **2018**, *30*, 509–527. [CrossRef]

45. Callioglu, H.; Topcu, M.; Tarakcılar, A.R. Elastic–plastic stress analysis of an orthotropic rotating disc. *Int. J. Mech. Sci.* **2006**, *48*, 985–990. [CrossRef]

46. Tarfaoui, M.; Nachtane, M.; Khadimallah, H.; Saifaoui, D. Simulation of mechanical behavior and damage of a large composite wind turbine blade under critical loads. *Appl. Compos. Mater.* **2018**, *25*, 237–254. [CrossRef]

47. Quadrino, A.; Penna, R.; Feo, L.; Nicola Nistico, N. Mechanical characterization of pultruded elements: Fiber orientation influence vs web-flange junction local problem. *Exp. Numer. Tests Compos. Part B* **2018**, *142*, 68–84. [CrossRef]

48. Morgado, T.; Silvestre, N.; Correia, J.R. Simulation of fire resistance behaviour of pultruded GFRP beams —Part II: Stress analysis and failure criteria. *Comp. Struct.* **2018**, *188*, 519–530. [CrossRef]

49. Zhou, Y.; Duan, M.; Ma, J.; Sun, G. Theoretical analysis of reinforcement layers in bonded flexible marine hose under internal pressure. *Eng. Struct.* **2018**, *168*, 384–398. [CrossRef]

Article

Compression of a Polar Orthotropic Wedge between Rotating Plates: Distinguished Features of the Solution

Sergei Alexandrov [1,2,*]**, Elena Lyamina** [2]**, Pham Chinh** [3] **and Lihui Lang** [1]

[1] School of Mechanical Engineering and Automation, Beihang University, Beijing 100191, China; lang@buaa.edu.cn

[2] Russian Academy of Sciences, Ishlinsky Institute for Problems in Mechanics RAS, 101-1 Prospect Vernadskogo, 119526 Moscow, Russia; lyamina@inbox.ru

[3] Vietnam Academy of Science and Technology, Institute of Mechanics, Hanoi 264 Doi Can, Vietnam; pdchinh@imech.vast.vn

* Correspondence: sergei_alexandrov@spartak.ru

Received: 29 January 2019; Accepted: 19 February 2019; Published: 20 February 2019

Abstract: An infinite wedge of orthotropic material is confined between two rotating planar rough plates, which are inclined at an angle 2α. An instantaneous boundary value problem for the flow of the material is formulated and solved for the stress and the velocity fields, the solution being in closed form. The solution may exhibit the regimes of sliding or sticking at the plates. It is shown that the overall structure of the solution significantly depends on the friction stress at sliding. This stress is postulated by the friction law. Solutions, which exhibit sticking, may exist only if the postulated friction stress at sliding satisfies a certain condition. These solutions have a rigid rotating zone in the region adjacent to the plates, unless the angle α is equal to a certain critical value. Solutions which exhibit sliding may be singular. In particular, some space stress and velocity derivatives approach infinity in the vicinity of the friction surface.

Keywords: polar orthotropy; Hill's yield criterion; friction regimes; singularity

---

1. Introduction

An instantaneous plane strain rigid plastic solution is obtained for compression of an infinite wedge of orthotropic material confined between two rough plates, inclined at angle 2α, and which intersect in a line. This boundary value problem is ideal for studying qualitative mathematical properties of boundary value problems, including constitutive equations and boundary conditions. For, exact analytical or semi-analytical solutions can be found for many constitutive equations. In particular, such solutions have been presented in [1,2] for isotropic viscoplastic materials and in [3] for the double slip and rotation model. A description of this model can be found in [4].

The present paper provides an analytic solution for rigid plastic orthotropic material. It is assumed that the principal axes of anisotropy are straight lines through the apex of the wedge and orthogonal curves, which are of course circular arcs. This type of orthotropy is of practical interest [5–8] among many others. The paper focuses on qualitative features of the solution such as non-existence of the solution, singularity in the stress and velocity fields, appearance of a rigid region near the plates and transition between the regimes of sticking and sliding. The effect of plastic anisotropy on these features is discussed.

The stress and velocity fields are singular if the regime of sliding occurs in the case of the maximum friction law. A detailed asymptotic analysis of the solution is performed for this case. In particular, it is shown that the asymptotic behavior of the solution is in agreement with the general theory developed in [9].

An applied aspect of the solution found, is that it can be used in conjunction with the method for analysis and the design of flat-rolling proposed in [10]. It is known that solutions, found by means of this method, show a good comparison with experiment [11–13], and are used for verifying solutions found by means of other approximate methods [14,15]. The importance of developing fast approximate methods for the analysis and design of the process of rolling has been emphasized in [16].

2. Statement of the Problem

Two semi-infinite rough plates rotate towards each other with angular velocity of magnitude ω about an axis O and compress a wedge of polar orthotropic material. The plates are inclined to each other at an angle 2α (Figure 1). The boundary value problem consists of the instantaneous plane strain deformation of the wedge. The problem is solved in a system of plane polar coordinates (r, θ) with its origin at O and with $\theta = 0$, taken as the perpendicular bisector of the angle 2α. It is assumed that the principal axes of anisotropy coincide with coordinate curves of the coordinate system chosen. Then, $\theta = 0$ is an axis of symmetry for the flow and it is sufficient to find the solution in the region $\theta \geq 0$. The components of the stress tensor referred to the polar coordinate system are denoted as σ_{rr}, $\sigma_{\theta\theta}$ and $\sigma_{r\theta}$; and the components of the velocity vector as u_r and u_θ. There is no material flux through O.

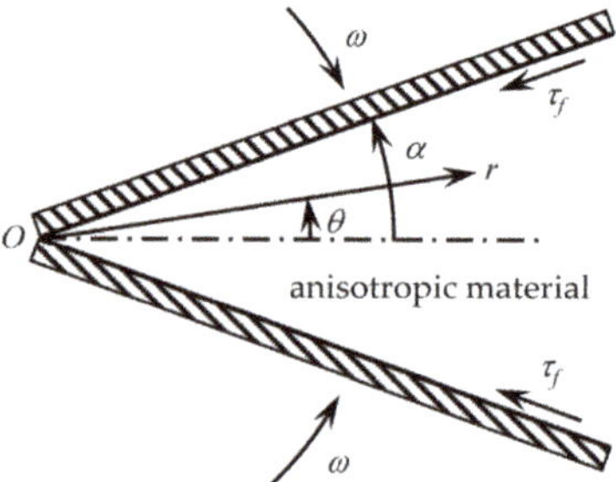

Figure 1. Geometry of the boundary value problem.

Therefore, the radial velocity should satisfy the following condition:

$$u_r = 0 \tag{1}$$

at $r = 0$. By symmetry,

$$u_\theta = 0 \tag{2}$$

and

$$\sigma_{r\theta} = 0 \tag{3}$$

at $\theta = 0$. The circumferential velocity should also satisfy the condition:

$$u_\theta = -\omega r \tag{4}$$

at $\theta = \alpha$. Finally, the friction law is taken in the form:

$$\begin{cases} u_r = 0 \, \text{if} \, |\sigma_{r\theta}| \leq \tau_f \\ \sigma_{r\theta} = -\tau_f \, \text{otherwise} \end{cases} \tag{5}$$

at $\theta = \alpha$. Here $\tau_f > 0$ denotes the frictional stress at sliding. The magnitude of τ_f will be specified later. The sense of $\sigma_{r\theta}$ in (5) is dictated by the condition that $u_r \geq 0$ at the plate.

It is assumed that the material obeys Hill's quadratic yield criterion [17] and its associated flow rule. The elastic portion of strain is neglected. In the case of plane strain deformation of a polar orthotropic material, whose principal axes of anisotropy coincide with the coordinate curves of the polar coordinate system, the constitutive equations of the model are:

$$\frac{(\sigma_{rr} - \sigma_{\theta\theta})^2}{4(1-c)} + \sigma_{r\theta}^2 = T^2 \tag{6}$$

and

$$\zeta_{rr} = \lambda\frac{(\sigma_{rr} - \sigma_{\theta\theta})}{2(1-c)}, \quad \zeta_{\theta\theta} = -\lambda\frac{(\sigma_{rr} - \sigma_{\theta\theta})}{2(1-c)}, \zeta_{r\theta} = \lambda\sigma_{r\theta} \tag{7}$$

where (6) in the yield criterion and (7) is the associated flow rule. The quantity T is the shear yield stress in the coordinate system chosen, c is a constitutive parameter, λ is a non-negative multiplier, ζ_{rr}, $\zeta_{\theta\theta}$, $\zeta_{r\theta}$ denote the components of the strain rate tensor. The parameter c can be expressed in terms of the yield stresses in the directions of the principal axes of anisotropy and can vary (theoretically) in the range $-\infty < c < 1$ [17]. Eliminating λ between the equations in (7) yields:

$$\zeta_{rr} + \zeta_{\theta\theta} = 0, \quad \frac{\zeta_{r\theta}}{\zeta_{rr} - \zeta_{\theta\theta}} = \frac{(1-c)\sigma_{r\theta}}{\sigma_{rr} - \sigma_{\theta\theta}}. \tag{8}$$

It is evident that the first equation here is the equation of incompressibility. The strain rate components are expressed in terms of the velocity components as

$$\zeta_{rr} = \frac{\partial u_r}{\partial r}, \quad \zeta_{\theta\theta} = \frac{1}{r}\frac{\partial u_\theta}{\partial \theta} + \frac{u_r}{r}, \quad \zeta_{r\theta} = \frac{1}{2}\left(\frac{1}{r}\frac{\partial u_r}{\partial \theta} + \frac{\partial u_\theta}{\partial r} - \frac{u_\theta}{r}\right). \tag{9}$$

The system of Equations (6), (8) and (9) are supplemented by the stress equilibrium equations:

$$\frac{\partial \sigma_{rr}}{\partial r} + \frac{1}{r}\frac{\partial \sigma_{r\theta}}{\partial \theta} + \frac{\sigma_{rr} - \sigma_{\theta\theta}}{r} = 0, \quad \frac{\partial \sigma_{r\theta}}{\partial r} + \frac{1}{r}\frac{\partial \sigma_{\theta\theta}}{\partial \theta} + \frac{2\sigma_{r\theta}}{r} = 0. \tag{10}$$

In total, there are five unknowns (three components of the stress tensor and two components of the velocity vector). The equations to solve are (6), (8) and (10). It is understood here that the components of the strain rate tensor in (8) should be eliminated by means of (9). The solution should satisfy the conditions (1) to (5).

3. General Stress Solution

The yield criterion (6) is satisfied by the following substitution:

$$\sigma_{rr} = \sigma + T\sqrt{1-c}\cos 2\varphi, \quad \sigma_{\theta\theta} = \sigma - T\sqrt{1-c}\cos 2\varphi, \quad \sigma_{r\theta} = -T\sin 2\varphi \tag{11}$$

where σ and φ are new unknown functions of r and θ. The direction of flow dictates that $\sigma_{r\theta} \leq 0$ and $\sigma_{rr} - \sigma_{\theta\theta} \geq 0$. Then, it is immediate from (11) that:

$$0 \leq \varphi \leq \frac{\pi}{4}. \tag{12}$$

Using (11) and (12) the boundary condition (3) transforms to:

$$\varphi = 0 \tag{13}$$

at $\theta = 0$. Substituting (11) into (10) gives:

$$\begin{aligned}
\frac{\partial \sigma}{\partial r} - 2T\sqrt{1-c}\sin 2\varphi\frac{\partial \varphi}{\partial r} - \frac{2T\cos 2\varphi}{r}\frac{\partial \varphi}{\partial \theta} + \frac{2T\sqrt{1-c}\cos 2\varphi}{r} &= 0, \\
-2T\cos 2\varphi\frac{\partial \varphi}{\partial r} + \frac{\partial \sigma}{r\partial \theta} - \frac{2T\sqrt{1-c}\sin 2\varphi}{r}\frac{\partial \varphi}{\partial \theta} - \frac{2T\sin 2\varphi}{r} &= 0.
\end{aligned} \tag{14}$$

A standard assumption made in similar problems of the classical theory of plasticity is that φ is independent of r [17]. In this case, the equations in (14) become:

$$\frac{\partial \sigma}{\partial r} - \frac{2T\cos 2\varphi}{r}\frac{d\varphi}{d\theta} + \frac{2T\sqrt{1-c}\cos 2\varphi}{r} = 0, \qquad \frac{\partial \sigma}{2T\partial\theta} - \sqrt{1-c}\sin 2\varphi\frac{d\varphi}{d\theta} - \sin 2\varphi = 0. \qquad (15)$$

The first equation can be immediately integrated to give:

$$\frac{\sigma}{2T} = \left(\frac{d\varphi}{d\theta} - \sqrt{1-c}\right)\cos 2\varphi \ln\left(\frac{r}{r_0}\right) + \frac{\sigma_0(\theta)}{2T}. \qquad (16)$$

Here r_0 is a constant introduced for convenience and $\sigma_0(\theta)$ is an arbitrary function of θ. Substituting (16) into the second equation in (15) yields:

$$\frac{d}{d\theta}\left[\left(\frac{d\varphi}{d\theta} - \sqrt{1-c}\right)\cos 2\varphi\right]\ln\left(\frac{r}{r_0}\right) = -\sqrt{1-c}\sin 2\varphi\frac{d\varphi}{d\theta} + \sin 2\varphi - \frac{d\sigma_0}{2Td\theta}. \qquad (17)$$

Since the right-hand side of this equation is independent of r, the coefficient of $\ln(r/r_0)$ on the left-hand side must vanish. Then, Equation (17) results in the following two equations:

$$\left(\frac{d\varphi}{d\theta} - \sqrt{1-c}\right)\cos 2\varphi = K_0\sqrt{1-c}, \qquad \frac{d\sigma_0}{2Td\theta} = \left(-\sqrt{1-c}\frac{d\varphi}{d\theta} + 1\right)\sin 2\varphi. \qquad (18)$$

Here K_0 is a constant of integration. Equation (16) becomes:

$$\frac{\sigma}{2T} = K_0\sqrt{1-c}\ln\left(\frac{r}{r_0}\right) + \frac{\sigma_0(\theta)}{2T}. \qquad (19)$$

The second equation in (18) can be rewritten as:

$$\frac{d\sigma_0}{2Td\varphi} = \left(-\sqrt{1-c} + \frac{d\theta}{d\varphi}\right)\sin 2\varphi.$$

Eliminating in this equation the derivative $d\theta/d\varphi$ by means of the first equation in (18) leads to:

$$\frac{d\sigma_0}{2Td\varphi} = \frac{1}{\sqrt{1-c}}\left[1 - c + \frac{\cos 2\varphi}{(K_0 + \cos 2\varphi)}\right]\sin 2\varphi.$$

Integrating gives:

$$\frac{\sigma_0}{2T} = \frac{1}{2\sqrt{1-c}}[-\cos 2\varphi + K_0 ln(K_0 + \cos 2\varphi) + K_1] \qquad (20)$$

where K_1 is constant of integration.

It is seen from (12) and (13) that $d\varphi/d\theta > 0$ at $\varphi = 0$. Therefore, it follows from the first equation in (18) that

$$K_0 > -1. \qquad (21)$$

The first equation in (18) can be integrated to give:

$$\theta\sqrt{1-c} = \varphi - K_0\text{arctanh}\left[\sqrt{\frac{1-K_0}{1+K_0}}\tan\varphi\right]\left(1 - K_0^2\right)^{-1/2} \qquad (22)$$

if $|K_0| < 1$,

$$\theta\sqrt{1-c} = \varphi - K_0\text{arctan}\left[\sqrt{\frac{K_0-1}{K_0+1}}\tan\varphi\right]\left(K_0^2 - 1\right)^{-1/2} \qquad (23)$$

if $K_0 > 1$, and

$$\theta\sqrt{1-c} = \varphi - \frac{\tan\varphi}{2} \tag{24}$$

if $K_0 = 1$. The solution for an important special case of (22), $K_0 = 0$, is represented as

$$\theta\sqrt{1-c} = \varphi. \tag{25}$$

The constant K_0 cannot be determined without the solution for velocity.

4. General Velocity Solution

The velocity components may be represented as:

$$u_r = \frac{\omega r}{2}\frac{dg(\varphi)}{d\theta} \quad\text{and}\quad u_\theta = -\omega r g(\varphi). \tag{26}$$

The condition (1) and the first equation in (8) are then automatically satisfied for any choice of the function $g(\varphi)$. Equations (9) and (26) combine to give:

$$\zeta_{rr} = \frac{\omega}{2}\frac{dg(\varphi)}{d\theta}, \quad \zeta_{\theta\theta} = -\frac{\omega}{2}\frac{dg(\varphi)}{d\theta}, \quad \zeta_{r\theta} = \frac{\omega}{4}\frac{d^2g(\varphi)}{d\theta^2}. \tag{27}$$

Substituting (11) and (27) in the second equation in (8) yields:

$$\frac{d^2g}{d\theta^2} + 2\sqrt{1-c}\tan 2\varphi\frac{dg}{d\theta} = 0 \tag{28}$$

or

$$\frac{dG}{d\theta} + 2\sqrt{1-c}\tan 2\varphi G = 0 \tag{29}$$

where $G = dg/d\theta$. Replacing in (29) differentiation with respect to θ with differentiation with respect to φ by means of the first equation in (18) results in:

$$\frac{dG}{d\varphi} = -\frac{2\sin 2\varphi}{(K_0 + \cos 2\varphi)}G. \tag{30}$$

Integrating gives

$$G = G_0(K_0 + \cos 2\varphi). \tag{31}$$

Here G_0 is constant of integration. The definition for G and (31) combine to give:

$$\frac{dg}{d\theta} = G_0(K_0 + \cos 2\varphi). \tag{32}$$

Replacing here differentiation with respect to θ with differentiation with respect to φ by means of the first equation in (18) results in:

$$\frac{dg}{d\varphi} = \frac{G_0\cos 2\varphi}{\sqrt{1-c}}. \tag{33}$$

It is seen from (2), (13) and (26) that $g = 0$ at $\varphi = 0$. The solution of Equation (32) satisfying this condition is:

$$g = \frac{G_0\sin 2\varphi}{2\sqrt{1-c}}. \tag{34}$$

Substituting (33) into (26) and then the resulting expression for the circumferential velocity into (4) yields:

$$G_0\sin 2\varphi_w = 2\sqrt{1-c} \tag{35}$$

where φ_w is the value of φ at $\theta = \alpha$. The dependence of φ_w on α follows from the solution of the first equation in (18).

To complete the solution of the boundary value problem, it is necessary to satisfy the boundary condition (5).

5. Solution of the Boundary Value Problem

The boundary condition (5) comprises two friction regimes, sticking and sliding. These regimes should be treated separately.

5.1. Regime of Sticking

In this regime, the boundary condition (5) becomes $u_r = 0$ at $\theta = \alpha$. It is seen from the definition for G and (26) that this condition is equivalent to the condition $G = 0$ at $\theta = \alpha$. Then, it follows from (31) that:

$$K_0 = -\cos 2\varphi_w. \tag{36}$$

In this case the dependence of θ on φ is given by (22). Eliminating K_0 in (22) by means of (36), it is possible to find that the argument of the inverse hyperbolic tangent function is equal to 1 at $\theta = \alpha$. Therefore, the left-hand side of (22) approaches infinity (or negative infinity) unless $K_0 = 0$. In the latter case, it is more convenient to use the solution (25). It follows from this solution, (35) and (36) that

$$\varphi_w = \frac{\pi}{4}, \quad G_0 = 2\sqrt{1-c} \text{ and } \alpha = \frac{\pi}{4\sqrt{1-c}} = \alpha_{cr}. \tag{37}$$

The solution at sticking is possible only if α and c satisfy the third equation. Another restriction on the existence of the solution at sticking is that the shear stress at $\theta = \alpha$ is less or equal to τ_f involved in (5). It is seen from the first equation in (37) and (11) that $\sigma_{r\theta} = -T$ at $\theta = \alpha$ if the regime of sticking occurs. Since T is the maximum possible value of the shear stress in the polar coordinate system, a necessary condition for the existence of the regime of sticking is that $\tau_f = T$. If $\tau_f < T$ then no solution at sticking exists.

Assume that $\tau_f = T$. The relation between α and c in (37) has been derived assuming that plastic yielding occurs in the region $0 \le \theta \le \alpha$. In the case of rigid/plastic solids, rigid regions may appear. In the case under consideration, the solution at sticking is possible if $\alpha > \alpha_{cr}$ and the material in the region $\alpha \ge \theta \ge \alpha_{cr}$ is rigid. It worthy of note that the stress solution at $K_0 = 0$ given in Section 3 is valid in the rigid region. Therefore, the yield criterion is not violated in the range $\alpha \ge \theta \ge \alpha_{cr}$ and the solution is complete.

5.2. Regime of Sliding

It is convenient to consider two cases, $\tau_f = T$ and $\tau_f < T$, separately. Assume that $\tau_f = T$ and $\alpha < \alpha_{cr}$. Then, no solution at sticking exists and it is necessary to find the solution at sliding. It follows from (5), (11) and (35) that the first and second equations in (37) are valid. The equation for determining K_0 follows from (22) or (23). It is however convenient to start with the special case $K_0 = 1$. In this case Equation (24) is valid. Therefore, this special case occurs only if α and c satisfy the following equation:

$$\alpha = \frac{(\pi - 2)}{4\sqrt{1-c}} = \alpha_s. \tag{38}$$

It is evident from (37) and (38) that $\alpha_s < \alpha_{cr}$. Equation (22) is valid in the range $\alpha_s < \alpha < \alpha_{cr}$. In this case, the equation for K_0 is

$$\alpha\sqrt{1-c} = \frac{\pi}{4} - K_0 \operatorname{arctanh}\left[\sqrt{\frac{1-K_0}{1+K_0}}\right]\left(1 - K_0^2\right)^{-1/2}. \tag{39}$$

Equation (23) is valid in the range $0 < \alpha < \alpha_s$. In this case, the equation for K_0 is

$$\alpha\sqrt{1-c} = \frac{\pi}{4} - K_0 \arctan\left[\sqrt{\frac{K_0-1}{K_0+1}}\right]\left(K_0^2 - 1\right)^{-1/2}. \tag{40}$$

Equations (39) and (40) should be solved numerically.

Prandtl's friction law reads $\tau_f = mT$ where $0 \le m \le 1$. The case $m = 1$ has been treated above. Therefore, assume that $m < 1$. In this case, no solution at sticking exists. The friction law (5) becomes $\sigma_{r\theta} = -mT$ at $\theta = \alpha$ (or $\varphi = \varphi_w$). Then, it follows from (11) that:

$$\varphi_w = \frac{1}{2}\arcsin m. \tag{41}$$

The value of G_0 is found from (35) and (41) as:

$$G_0 = \frac{2\sqrt{1-c}}{m}. \tag{42}$$

The equation for determining K_0 follows from (22) or (23). As before, it is more convenient to consider special cases first. The values of α_{cr} and α_s are now determined from (24) and (25) as:

$$\alpha_{cr} = \frac{\varphi_w}{\sqrt{1-c}} \quad \text{and} \quad \alpha_s = \left(\varphi_w - \frac{\tan\varphi_w}{2}\right)\frac{1}{\sqrt{1-c}}. \tag{43}$$

In these equations, φ_w should be eliminated by means of (41). Equation (22) is valid in the ranges $\alpha_s < \alpha < \alpha_{cr}$ and $\alpha > \alpha_{cr}$. In this case, the equation for K_0 is:

$$\alpha\sqrt{1-c} = \varphi_w - K_0 \operatorname{arctanh}\left[\sqrt{\frac{1-K_0}{1+K_0}}\tan\varphi_w\right]\left(1 - K_0^2\right)^{-1/2}. \tag{44}$$

The value of K_0 is positive in the range $\alpha_s < \alpha < \alpha_{cr}$ and negative in the range $\alpha > \alpha_{cr}$. Equation (23) is valid in the ranges $0 < \alpha < \alpha_s$. In this case, the equation for K_0 is:

$$\alpha\sqrt{1-c} = \varphi_w - K_0 \arctan\left[\sqrt{\frac{K_0-1}{K_0+1}}\tan\varphi_w\right]\left(K_0^2 - 1\right)^{-1/2}. \tag{45}$$

6. Singularity

It is seen from (18) that the derivative $d\varphi/d\theta$ approaches infinity as $\varphi \to \pi/4$ if $K_0 \neq 0$. If $m < 1$ then $\varphi_w < \pi/4$ and the solution is not singular. If the regime of sticking occurs then $\varphi_w = \pi/4$ but $K_0 = 0$. Therefore, the solution may be singular only if $m = 1$ and the regime of sliding occurs. It follows from (18) that:

$$\frac{d\varphi}{d\theta} = \frac{K_0\sqrt{1-c}}{2(\pi/4 - \varphi)} + O(1) \tag{46}$$

as $\varphi \to \pi/4$. Integrating and using the boundary condition $\varphi = \pi/4$ at $\theta = \alpha$ yields:

$$\frac{\pi}{4} - \varphi = \sqrt{K_0\sqrt{1-c}}\sqrt{\theta - \alpha} + o\left(\sqrt{\theta - \alpha}\right) \tag{47}$$

as $\theta \to \alpha$.

Consider the stress field. Differentiating (11) with respect to θ yields:

$$\frac{\partial\sigma_{rr}}{\partial\theta} = \frac{\partial\sigma}{\partial\theta} - 2T\sqrt{1-c}\sin 2\varphi\frac{d\varphi}{d\theta}, \quad \frac{\partial\sigma_{\theta\theta}}{\partial\theta} = \frac{\partial\sigma}{\partial\theta} + 2T\sqrt{1-c}\sin 2\varphi\frac{d\varphi}{d\theta}, \quad \frac{\partial\sigma_{r\theta}}{\partial\theta} = -2T\cos 2\varphi\frac{d\varphi}{d\theta}. \tag{48}$$

Eliminating the derivative $d\varphi/d\theta$ in these equations by means of (18) gives:

$$\frac{\partial\sigma_{rr}}{\partial\theta} = \frac{\partial\sigma}{\partial\theta} - \frac{2T(1-c)(K_0+\cos 2\varphi)\sin 2\varphi}{\cos 2\varphi}, \quad \frac{\partial\sigma_{\theta\theta}}{\partial\theta} = \frac{\partial\sigma}{\partial\theta} + \frac{2T(1-c)(K_0+\cos 2\varphi)\sin 2\varphi}{\cos 2\varphi},$$
$$\frac{\partial\sigma_{r\theta}}{\partial\theta} = -2T\sqrt{1-c}(K_0+\cos 2\varphi). \tag{49}$$

It is evident that the derivative $\partial\sigma_{r\theta}/\partial\theta$ is of a finite magnitude at $\varphi = \pi/4$ (or $\theta = \alpha$). The derivative $\partial\sigma/\partial\theta$ involved in (49) is determined from (18), (19) and (20) as:

$$\frac{\partial\sigma}{\partial\theta} = 2T\left[1 - \frac{(1-c)(K_0+\cos 2\varphi)}{\cos 2\varphi}\right]\sin 2\varphi. \tag{50}$$

Equations (49) and (50) combine to give:

$$\frac{\partial\sigma_{rr}}{\partial\theta} = 2T\left[1 - \frac{2(1-c)(K_0+\cos 2\varphi)}{\cos 2\varphi}\right]\sin 2\varphi, \quad \frac{\partial\sigma_{\theta\theta}}{\partial\theta} = 2T\sin 2\varphi. \tag{51}$$

It is evident that the derivative $\partial\sigma_{\theta\theta}/\partial\theta$ is of a finite magnitude at $\varphi = \pi/4$ (or $\theta = \alpha$). Expanding the right-hand side of the first equation in (51) in a series in the vicinity of $\varphi = \pi/4$ results in

$$\frac{\partial\sigma_{rr}}{\partial\theta} = -\frac{2TK_0(1-c)}{(\pi/4-\varphi)} + o\left[(\pi/4-\varphi)^{-1}\right] \tag{52}$$

as $\varphi \to \pi/4$. Equations (47) and (52) combine to give:

$$\frac{\partial\sigma_{rr}}{\partial\theta} = -\frac{2TK_0(1-c)}{\sqrt{K_0}\sqrt{1-c}\sqrt{\theta-\alpha}} + o\left[(\theta-\alpha)^{-1/2}\right] \tag{53}$$

as $\theta \to \alpha$. It is seen from this equation that the derivative $\partial\sigma_{rr}/\partial\theta$ approaches infinity (or negative infinity) in the vicinity of the friction surface and follows an inverse square root rule.

Consider the strain rate field. It follows from the definition for G, (27) and (31), that $\xi_{rr} = -\xi_{\theta\theta} = \omega G_0(K_0+\cos 2\varphi)/2$. It is evident from this equation that the normal strain rates in the polar coordinate system are bounded at the friction surface. The shear strain rate is determined from (18), (27) and (31) as:

$$\xi_{r\theta} = -\frac{\omega G_0\sqrt{1-c}(K_0+\cos 2\varphi)\tan 2\varphi}{2}. \tag{54}$$

It is seen from this equation that $|\xi_{r\theta}| \to \infty$ as $\varphi \to \pi/4$. Expanding the right-hand side of (54) in a series in the vicinity of $\varphi = \pi/4$ results in:

$$\xi_{r\theta} = -\frac{\omega G_0 K_0\sqrt{1-c}}{4}\left(\frac{\pi}{4}-\varphi\right)^{-1} + o\left[\left(\frac{\pi}{4}-\varphi\right)^{-1}\right] \tag{55}$$

as $\varphi \to \pi/4$. Equations (47) and (55) combine to give:

$$\xi_{r\theta} = -\frac{\omega G_0\sqrt{K_0}\sqrt{1-c}}{4\sqrt{\theta-\alpha}} + o\left[(\theta-\alpha)^{-1/2}\right] \tag{56}$$

as $\theta \to \alpha$. It is seen from this equation that the shear strain rate in the polar coordinate system follows an inverse square root rule in the vicinity of the friction surface. This result is in agreement with the general theory developed in [9].

Some models of anisotropic plasticity (for example, [18]) involve the material spin. Therefore, it is of interest to understand the asymptotic behavior of the only non-zero spin component, $\omega_{r\theta}$, near the friction surface. By definition,

$$\omega_{r\theta} = \frac{1}{2}\left(\frac{1}{r}\frac{\partial u_r}{\partial\theta} - \frac{\partial u_\theta}{\partial r} - \frac{u_\theta}{r}\right). \tag{57}$$

Equations (26) and (57) combine to give:

$$\omega_{r\theta} = \frac{\omega}{4}\left(\frac{d^2 g}{d\theta^2} + 4g\right).$$

(58)

Using the definition for G, (18), (31) and (34) Equation (58) can be rewritten as:

$$\omega_{r\theta} = \frac{\omega G_0 \sin 2\varphi}{4}\left[\frac{1}{\sqrt{1-c}} - \frac{2\sqrt{1-c}}{\cos 2\varphi}(K_0 + \cos 2\varphi)\right].$$

(59)

It is seen from this equation that $|\omega_{r\theta}| \to \infty$ as $\varphi \to \pi/4$. Expanding the right-hand side of (59) in a series in the vicinity of $\varphi = \pi/4$ results in:

$$\omega_{r\theta} = -\frac{\omega\sqrt{1-c}G_0 K_0}{4}\left(\frac{\pi}{4} - \varphi\right)^{-1} + o\left[\left(\frac{\pi}{4} - \varphi\right)^{-1}\right]$$

(60)

as $\varphi \to \pi/4$. Equations (47) and (60) combine to give:

$$\omega_{r\theta} = -\frac{\omega\sqrt{\sqrt{1-c}K_0 G_0}}{4\sqrt{\theta - \alpha}} + o\left[(\theta - \alpha)^{-1/2}\right]$$

(61)

as $\theta \to \alpha$. The qualitative behavior of the material spin near the friction surface that its magnitude approaches infinity should be taken into account in material models that involve this quantity. A similar approach has been used in visco-plasticity [19], where the qualitative behavior of the quadratic invariant of the strain tensor near the friction surface, that its magnitude approaches infinity has been taken into account.

7. Conclusions

The boundary value problem for the flow of the orthotropic material, resulting from the problem formulated in Section 2 and illustrated in Figure 1, has been solved with the resulting solution being in closed form. The stress field has been determined up to an arbitrary constant (K_1 in Equation (20)). Emphasized are the qualitative features of the solution. In particular, if the friction law demands that the friction stress at sliding is less than the shear yield stress referred to in the principal axes of anisotropy then:

1. no solution at sticking exists; and
2. the solution at sliding involves no rigid region.

If the friction law demands that the friction stress at sliding is equal to the shear yield stress referred to the principal axes of anisotropy then:

1. no solution at sticking exists if $\alpha < \alpha_{cr}$ (α_{cr} is introduced in (37)) and the solution for $\alpha > \alpha_{cr}$ requires a rigid region adjacent to the plate; and
2. the solution at sliding exists if $\alpha < \alpha_{cr}$ and this solution is singular (some stress and velocity derivatives approach infinity in the vicinity of the friction surface).

The effect of plastic anisotropy on the solution is controlled by the constitutive parameter c and $c = 0$ for isotropic material. Even though the qualitative features of the solution are independent of the value of c, the quantitative effect may be quite significant. For example, the values of two critical angles, α_{cr} and α_s (α_s is introduced in (38)), are sensitive to the value of c, and these angles control the overall structure of the solution.

Author Contributions: Conceptualization, S.A.; closed form solution, E.L.; statement of the boundary value problem P.C., project administration

Funding: This research was funded by the Russian Foundation for Basic Research (Project RFBR-19-51-52003) and Vietnam Academy of Science and Technology (Project QTRU01.05/18-19).

Acknowledgments: EL and PC acknowledge support from grants RFBR-19-51-52003 (Russia) and QTRU01.05/18-19 (Viet Nam).

Conflicts of Interest: The authors declare no conflict of interest.

References

1. Alexandrov, S.; Jeng, Y.-R. Compression of viscoplastic material between rotating plates. *Trans. ASME J. Appl. Mech.* **2009**, *76*. [CrossRef]
2. Alexandrov, S.; Miszuris, W. The transition of qualitative behaviour between rigid perfectly plastic and viscoplastic solutions. *J. Eng. Math.* **2016**, *97*, 67–81. [CrossRef]
3. Alexandrov, S.; Harris, D. An Exact solution for a model of pressure-dependent plasticity in an un-steady plane strain process. *Eur. J. Mech. -A/Solids* **2010**, *29*, 966–975. [CrossRef]
4. Harris, D.; Grekova, E.F. A hyperbolic well-posed model for the flow of granular materials. *J. Eng. Math.* **2005**, *52*, 107–135. [CrossRef]
5. Liang, D.S.; Wang, H.J.; Chen, L.W. Vibration and stability of rotating polar orthotropic annular disks subjected to a stationary concentrated transverse load. *J. Sound Vib.* **2002**, *250*, 795–811. [CrossRef]
6. Koo, K.N. Vibration analysis and critical speeds of polar orthotropic annular disks in rotation. *Compos. Struct.* **2006**, *76*, 67–72. [CrossRef]
7. Peng, X.-L.; Li, X.-F. Elastic analysis of rotating functionally graded polar orthotropic disks. *Int. J. Mech. Sci.* **2012**, *60*, 84–91. [CrossRef]
8. Jeong, W.; Alexandrov, S.; Lang, L. Effect of plastic anisotropy on the distribution of residual stresses and strains in rotating annular disks. *Symmetry* **2018**, *10*, 420. [CrossRef]
9. Alexandrov, S.; Jeng, Y.-R. Singular rigid/plastic solutions in anisotropic plasticity under plane strain conditions. *Cont. Mech. Therm.* **2013**, *25*, 685–689. [CrossRef]
10. Orowan, E. The calculation of roll pressure in hot and cold flat rolling. *Proc. Inst. Mech. Eng.* **1943**, *150*, 140–167. [CrossRef]
11. Kimura, H. Application of Orowan theory to hot rolling of aluminum. *J. Jpn. Inst. Light Met.* **1985**, *35*, 222–227. [CrossRef]
12. Kimura, H. Application of Orowan theory to hot rolling of aluminum: computer control of hot rolling of aluminum. *Sumitomo Light Met. Tech. Rep.* **1985**, *26*, 189–194.
13. Lenard, J.G.; Wang, F.; Nadkarni, G. Role of constitutive formulation in the analysis of hot rolling. *Trans. ASME J. Eng. Mater. Technol.* **1987**, *109*, 343–349. [CrossRef]
14. Atreya, A.; Lenard, J.G. Study of cold strip rolling. *Trans. ASME J. Eng. Mater. Technol.* **1979**, *101*, 129–134. [CrossRef]
15. Domanti, S.; McElwain, D.L.S. Two-dimensional plane strain rolling: an asymptotic approach to the estimation of inhomogeneous effects. *Int. J. Mech. Sci.* **1995**, *37*, 175–196. [CrossRef]
16. Cawthorn, C.J.; Loukaides, F.G.; Allwood, J.M. Comparison of analytical models for sheet rolling. *Proc. Eng.* **2014**, *81*, 2451–2456. [CrossRef]
17. Hill, R. *The Mathematical Theory of Plasticity*; Clarendon Press: Oxford, UK, 1950.
18. Collins, I.F.; Meguid, S.A. On the influence of hardening and anisotropy on the plane-strain compression of thin metal strip. *Trans. ASME J. Appl. Mech.* **1977**, *44*, 272–278. [CrossRef]
19. Alexandrov, S.; Mishuris, G. Viscoplasticity with a saturation stress: distinguished features of the model. *Arch. Appl. Mech.* **2007**, *77*, 35–47. [CrossRef]

Article

Generation of Numerical Models of Anisotropic Columnar Jointed Rock Mass Using Modified Centroidal Voronoi Diagrams

Qingxiang Meng [1,2,*], Long Yan [2,3,*], Yulong Chen [4,*] and Qiang Zhang [5,6]

1   College of Water Conservancy and Hydropower Engineering, Hohai University, Nanjing 210098, China
2   Research Institute of Geotechnical Engineering, Hohai University, Nanjing 210098, China
3   Key Laboratory of Ministry of Education for Geomechanics and Embankment Engineering, Hohai University, Nanjing 210098, China
4   Department of Hydraulic Engineering, Tsinghua University, Beijing 100084, China
5   China Institute of Water Resources and Hydropower Research, Beijing 100048, China; zhangq@iwhr.com
6   State Key Laboratory of Hydraulics and Mountain River Engineering, Sichuan University, Chengdu 610065, China
*   Correspondence: tianyameng@hhu.edu.cn (Q.M.); loongyan@outlook.com (L.Y.); chen_yl@tsinghua.edu.cn (Y.C.)

Received: 21 October 2018; Accepted: 7 November 2018; Published: 9 November 2018

Abstract: A columnar joint network is a natural fracture pattern with high symmetry, which leads to the anisotropy mechanical property of columnar basalt. For a better understanding the mechanical behavior, a novel modeling method for columnar jointed rock mass through field investigation is proposed in this paper. Natural columnar jointed networks lies between random and centroidal Voronoi tessellations. This heterogeneity of columnar cells in shape and area can be represented using the coefficient of variation, which can be easily estimated. Using the bisection method, a modified Lloyd's algorithm is proposed to generate a Voronoi diagram with a specified coefficient of variation. Modelling of the columnar jointed rock mass using six parameters is then presented. A case study of columnar basalt at Baihetan Dam is performed to demonstrate the feasibility of this method. The results show that this method is applicable in the modeling of columnar jointed rock mass as well as similar polycrystalline materials.

Keywords: anisotropic columnar jointed rock; numerical model; centroidal Voronoi diagram; coefficient of variation

---

1. Introduction

Columnar jointed rock is a typical geological structure formed by a lot of ordered colonnades like the world famous natural wonders of Fingal's Cave and Giant's Causeway [1,2]. Being a miraculous natural phenomenon, the study on the geological structure of columnar jointed rock can be dated back to the 17th century [3]. Nowadays, the formation of columnar jointing is reasonably understood as a result of cracks propagation in cooling lava flows [4–7]. In situ observations and laboratory tests have also been employed in the study of columnar jointing [8–10].

With the increase of human activities, some engineering projects have encountered columnar jointing, and a large hydropower station is even founded on it [11]. Because of its adverse geologic conditions, study on the mechanical properties of columnar jointed rock mass is very important in engineering projects. King et al. studied elastic-wave propagation in columnar joints with a series of cross-hole acoustic measurements made between four horizontal boreholes drilled from a near-surface underground opening situated in a basaltic rock mass [12]. In connection with the Baihetan Hydropower Station project, several studies have been implemented for the mechanical property of columnar jointed rock. Meng systematically analyzed the anisotropic properties of columnar jointed basalt using analytical and numerical methods [13]. Zheng et al. proposed a 3-D modeling method for columnar basalt using random Voronoi tessellation [14]. In References [15–17], 2-D and 3-D discrete element simulation methods of columnar are implemented, and the representative elementary volume scale and equivalent mechanical parameters are obtained. Based on meso-structural analysis, macro-anisotropic constitutive model is developed and 3-D numerical simulation is performed for the dam foundation considering the effect of columnar joints [18]. Laboratory tests are carried out using experimental analogs of jointing under uniaxial compression [19]. These research results indicate that the characteristics of columnar jointed rock mass are very complex with anisotropic and nonlinear properties.

Most of the modelling methods of columnar jointed rock mass use the program developed by Zheng et al. [14]; however, this method has some shortcomings, such as a low efficiency and immutable columnar shapes, as pointed out in Reference [20]. The objective of this paper is to overcome these shortcomings in the generation of columnar joints in rock mass. An efficient and controllable method for the generation of columnar joints is proposed based on a constrained centroidal Voronoi tessellation diagram. In Section 2, the characteristics of typical columnar rock mass are briefly introduced. The properties of Voronoi diagrams and the latest related research on this topic are discussed in Section 3. Based on these discussion and algorithms, a detailed procedure for the generation of columnar jointed rock mass is developed in Section 4. A case study on columnar jointed basalt generation for the Baihetan Hydropower Station is presented in Section 5. Finally, the conclusion is given in Section 6.

2. Typical Shapes of Columnar Jointed Rock Mass

With the development of hydropower station, columnar jointing is often encountered in southwestern China. As a result of cooling lava flows and ash-flow tuffs, it occurs in many types of volcanic rocks in the formation of a regular array of polygonal prisms or columns [21]. Several famous typical columnar joints are shown in Figure 1.

(a)   (b)   (c)

Figure 1. Typical columnar jointed rock masses: (**a**) Devil's Postpile, Yosemite in California; (**b**) Fingal's Cave, Staffa in Scotland; and (**c**) Giant's Causeway, Antrim in Northern Ireland.

Although the characteristics of columnar jointed rock is typical, they vary for different locations. The column in some areas is regular and straight like Giant's Causeway, while some other places like Baihetan Dam, it is irregular. Besides, the diameters of columns vary from meters to centimeters. Columns are usually parallel and straight, and the length can be as much as several meters. Furthermore, the number of sides of an individual column also varies from three to eight. In order to give a more specific illustration of the columns, a colorized O'Reilly map is presented in Figure 2. In this figure, hexagons are green, heptagons are blue, octagons are purple, pentagons are orange, and squares are red. From the introduction of columnar jointed rock mass above, a diagrammatic drawing is proposed in Figure 3. It is appropriate to model columnar joints using Voronoi diagrams.

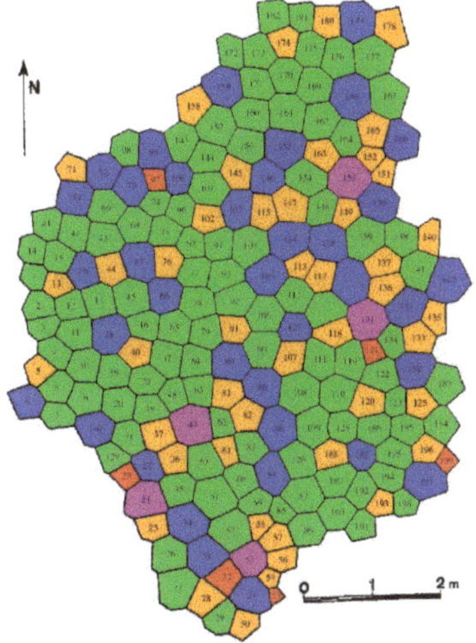

Figure 2. A colorized map of about 200 columns at the Giant's Causeway made by O'Reilly in 1879 [1,21].

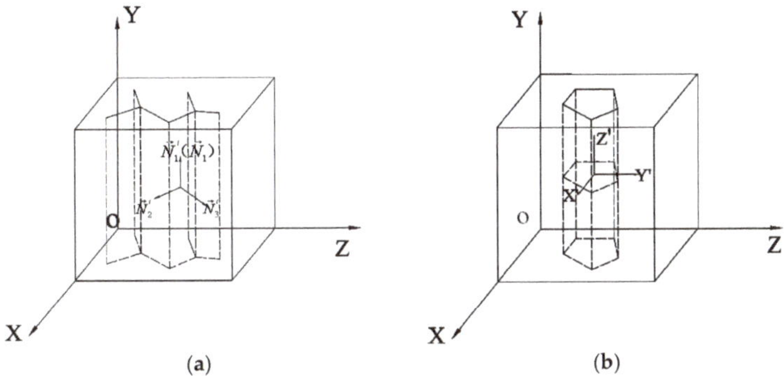

Figure 3. Diagrammatic drawing of the joints in columnar basalt: (**a**) vertical columnar joint, and (**b**) horizontal transverse joint.

3. Constrained Centroidal Voronoi and Implementation Method

3.1. Voronoi Diagram Algorithm and Its Constraints

3.1.1. Classical Voronoi Tessellation

A Voronoi diagram is a classical domain partitioning method named after Georgy Voronoi, a Russian mathematician. Voronoi generation is the dual algorithm of Delaunay triangulation. A typical Voronoi figure is generated using the perpendicular bisector of the lines composed by a set of points called seeds. The Voronoi diagram is a topic in computational geometry and has already widely applied in some other research areas like hydrology and crystal mechanics.

As shown in Figure 4, a Voronoi diagram with 10 random seeds P_i is generated. It can be seen that the domain is divided into 10 patches and each patch has a single seed. For the generation of a Voronoi diagram, a Delaunay triangulation is first implemented. Then the perpendicular bisector is plotted to partition the domain into different Voronoi cells. With the development of computational geometry, Voronoi tessellation is included in many software and packages like MATLAB, Mathematica, and SciPy. However, there are two obvious shortcomings of the classical Voronoi diagram:

(a) The Voronoi cell is not closed. For the Voronoi diagram in Figure 4, only one cell is closed and the other nine cells are open. This brings inconvenience for the analysis.

(b) The shape of the Voronoi cell is random and it is very hard to generate a Voronoi diagram with a specified statistical distribution.

3.1.2. Constrained Voronoi Diagram Generation

To overcome the first drawback of classical Voronoi tessellation, a constrained Voronoi diagram algorithm is employed to generate a closed Voronoi tessellation with seed points P_i and domain D. Taking a model with 16 points (x_i, y_i) and a square domain in Figure 5a as example, this algorithm can be described as follows [22]:

(a) A Voronoi diagram is generated using a classical tessellation method (Figure 5b).

(b) All the open cells with a vertex outside the domain D are identified (shown in Figure 5c).

(c) A set of new seeds symmetric to the seeds of open cells with respect to the domain boundary are created (Figure 5d).

(d) New Voronoi diagram is generated with a classical tessellation (Figure 5e).

(e) By removing the open cells as well as the related seeds, the final Voronoi diagram is shown in Figure 5f.

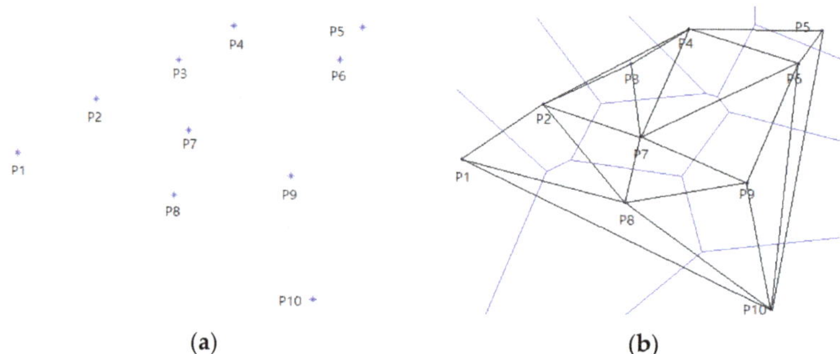

(a)   (b)

Figure 4. An illustration of Voronoi tessellation with 10 generators: (**a**) a set of generators, and (**b**) the resulting Voronoi diagram.

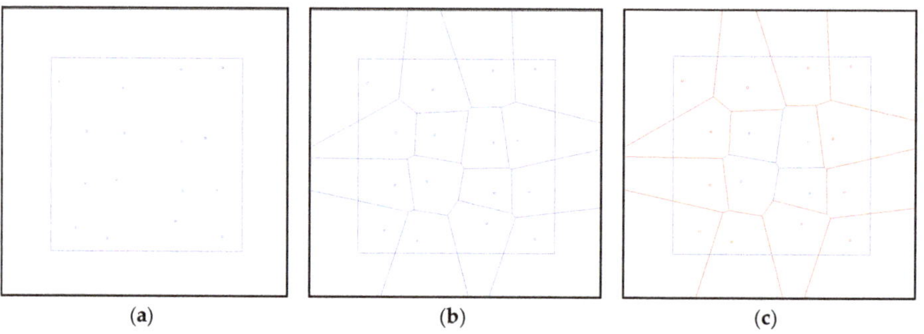

(a)   (b)   (c)

Figure 5. *Cont.*

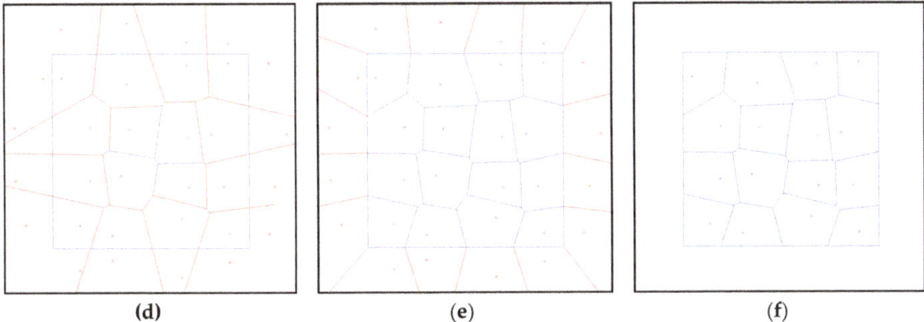

(d)   (e)   (f)

Figure 5. Illustration of the stages of the constrained Voronoi tessellation algorithm: (**a**) initial seeds, (**b**) classical Voronoi tessellation, (**c**) open Voronoi cells, (**d**) symmetry seeds, (**e**) new Voronoi tessellation, (**f**) constrained Voronoi diagram.

3.2. Centroidal Voronoi Algorithm

3.2.1. Random and Centroidal Voronoi Diagram

The Voronoi diagram created using a constrained Voronoi tessellation method with randomly generated seeds is shown in Figure 6a. It can be seen that the Voronoi cell is irregular and there is a clear deviation between the Voronoi seeds and the centroids. In order to obtain more regular columnar as shown in Figure 2, an algorithm for centroidal Voronoi diagrams (Figure 6b) is proposed to tackle this problem.

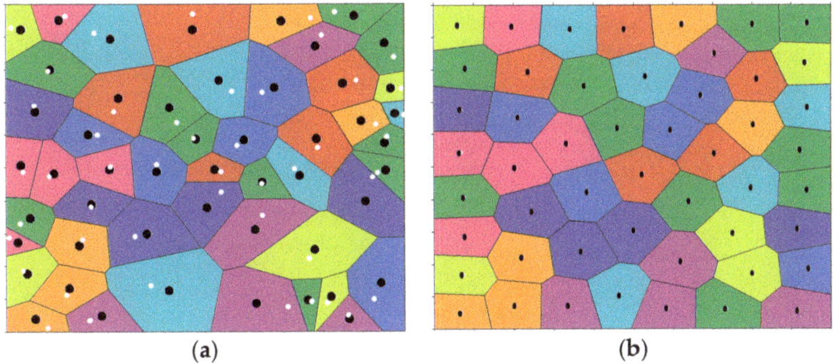

(a)   (b)

Figure 6. Random and centroidal Voronoi diagram: generators (white dots) and centroids (black dots): (**a**) random Voronoi diagram, and (**b**) centroidal Voronoi diagram.

Compared with the random Voronoi diagram, the seed of a cell is coincident with its centroid in centroidal Voronoi tessellation (CVT). It is widely used in related fields like mesh generation and data compression. Previous study has indicated that some natural patterns like Giant's Causeway can be represented by the centroidal Voronoi tessellation [5].

There are a number of methods for the generation of centroidal Voronoi diagram, among which are the algorithms by Lloyd and by MacQueen [23,24], which are widely used, and some other methods are variations of these two methods. Considering that MacQueen's algorithm needs twice as many Monte Carlo simulations in each iteration, which consumes a large amount computation time, Lloyd's centroidal method is employed for modelling columnar jointed rock mass in this study.

3.2.2. Lloyd's Algorithm

In this work, Lloyd's algorithm, named after Stuart P. Lloyd, is employed. It is an iteration method to partition the domain into well-shaped and uniformly-sized convex cells [25]. The convergence of Lloyd's algorithm to a centroidal Voronoi diagram has been proven. The random Voronoi diagram can be iterated to a centroidal Voronoi diagram using Lloyd's algorithm and the algorithm can be simplified as follows:

(1) For an initial seeds y_i, generate a Voronoi diagram using constrained Voronoi tessellation;
(2) Compute the centroid z_i of the Voronoi diagram of y_i;
(3) Move the generating point y_i to its centroid z_i;
(4) Repeat Steps 1 to 3 until all generating points converge to the centroids.

3.2.3. Estimation of the Centroid

The centroidal Voronoi algorithm is very simple but the calculation of the centroid of each Voronoi cell is time-consuming. In each iteration of Lloyd's method, the centroidal positions of all shapes of the Voronoi diagram are calculated. Thus, the estimation of the centroid is the most time-consuming task. For determining the centroid of a Voronoi region, a simple formula is:

$$z_i = \frac{\int_{V_i} \chi \rho(\chi) dV_i}{\int_{V_i} \rho(\chi) dV_i} \tag{1}$$

where χ is the position, V_i is the region area, and $\rho(\chi)$ is the density function with $\rho(\chi) = 1$ being the default.

In order to calculate the position of centroid fast, two efficient methods for evaluating the integrals are used in this work. One is based on an integration algorithm on triangle partitions and the other is a sampling method. An integration algorithm can calculate the coordinates of centroids exactly but not as fast as a sampling method; whereas a sampling method is fast but the results are not as accurate as that of an integration algorithm.

(1) Integration method on triangle partitions

For an arbitrary polygon with n vertices, it can be divided into $n - 2$ triangles with a clockwise or anti-clockwise direction (Figure 7). For each triangle with vertex A_i (x_i, y_i) ($i = 1, 2, 3$), the coordinates of centroid are given by:

$$x_g = (x_1 + x_2 + x_3)/3 \tag{2}$$

$$y_g = (y_1 + y_2 + y_3)/3 \tag{3}$$

The area of the triangle is calculated as:

$$S = ((x_2 - x_1) \times (y_3 - y_1) - (x_3 - x_1) \times (y_2 - y_1))/2 \tag{4}$$

Considering that the polygon can be discretized into $n - 2$ triangles and the centroid and area of each triangle can be expressed as $G_i(x_{gi}, y_{gi})$ and S_i, the coordinates of the centroid of the polygon can be obtained as:

$$X_g = \frac{\sum\limits_{i=1}^{n-2} x_{gi} S_i}{\sum\limits_{i=1}^{n-2} S_i} \tag{5}$$

$$Y_g = \frac{\sum_{i=1}^{n-2} y_{gi}S_i}{\sum_{i=1}^{n-2} S_i} \tag{6}$$

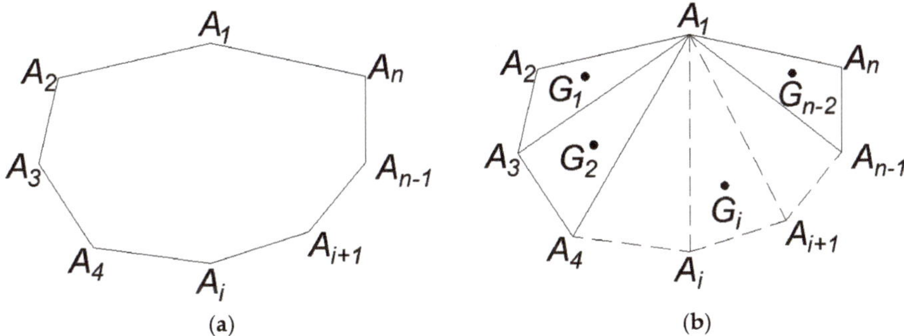

Figure 7. Discretization of n-polygon: (**a**) polygon with n vertices, and (**b**) $n-2$ discrete triangles.

(2) Sampling method

Although the integration on triangle partitions is a good solution for the estimation of the centroid of a polygon, it takes a long time to complete each step in Lloyd's iteration. For a quick calculation, a simple and efficient sampling method is proposed as an alternative.

For a certain area with different Voronoi tessellations, a set of random points are distributed in this area (Figure 8). For each Voronoi cell, if there are n points $P_i(x_i, y_i)$ in this area, then the centroid of this cell can be approximated as the average of these points:

$$X_g = \sum_{i=1}^{n} x_i/n \tag{7}$$

$$Y_g = \sum_{i=1}^{n} y_i/n \tag{8}$$

In this way, the determination of centroids changes to the partition of areas into different Voronoi cells, i.e., the "N-D nearest point search" problem. Luckily, the nearest point search problem has a mature mathematical solution [26] and this algorithm has been written in a number of software libraries, such as Python and MATLAB. Using these libraries, all the random points can be divided efficiently and the centroid of Voronoi diagram can be computed quickly. With more sampling points, the calculation of centroid will be more accurate. If the number of random points is selected properly, this algorithm can be both fast and relatively accurate.

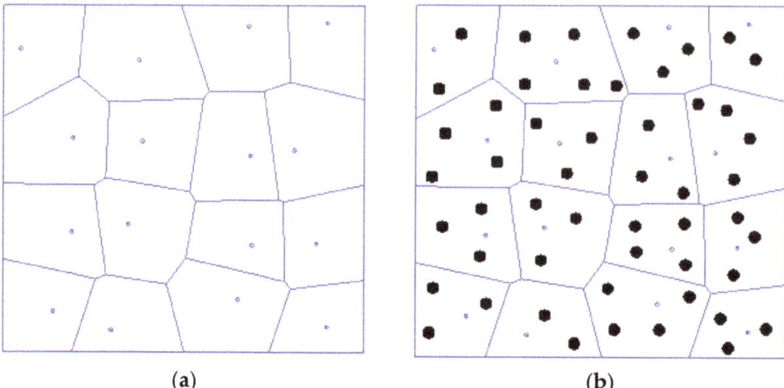

Figure 8. Illustration of the sampling method: (**a**) Voronoi tessellation, and (**b**) random sampling points.

3.3. Numerical Implementation and Discussion

In the iterations of Lloyd's algorithm, the Voronoi generators and centroids are known, and energy can be employed to describe the Voronoi diagram [23]. For the Voronoi tessellation $\{\Omega_i\}_{i=1}^{n}$, the energy is defined as:

$$E = \sum_{i=1}^{n} \int_{\Omega i} \rho(\chi)\| \chi - \chi_g \|^2 d\sigma \tag{9}$$

where χ is the position of the Voronoi generator, χ_g represents the position of the centroid, and $\rho(\chi) = 1$ is the density function with $\rho(\chi) = 1$ being the default.

However, if the generator of Voronoi diagram is not known beforehand, energy cannot be estimated. For example, it is not easy to calculate the energy of Voronoi cells in Figure 2. Indirectly, the coefficent of variation, which has been proven as an effective parameter in the estimation of Voronoi diagram [27,28], is employed for the description of the properties of Voronoi cells. In the process of smoothing energy from a randomly-generated Voronoi diagram to a centroidal Voronoi diagram, the coefficient of variation value is recorded with the following formula:

$$CV = \frac{SD}{m} \tag{10}$$

where SD is the standard deviation and m is the mean of Voronoi cell areas. The coefficient of variation gives a quantitative indication on the spatial distribution: the higher the coefficient of variation, the higher the tendency of cells to aggregate into clusters.

Based on the related algorithms described, numerical implementation is done in MATLAB, in which some good programming techniques from the open-source program of Burkardt are referenced [29]. The 50 seeds for Voronoi diagram are generated randomly. After 50 iterations, the points are well-spaced and Lloyd's algorithm for computing centroidal Voronoi tessellation converged overall (Figure 9). In this way, the coefficient of variation has the same evolution trend as energy (E), and it is reasonable to use coefficient of variation to describe the distribution property of the Voronoi tessellation.

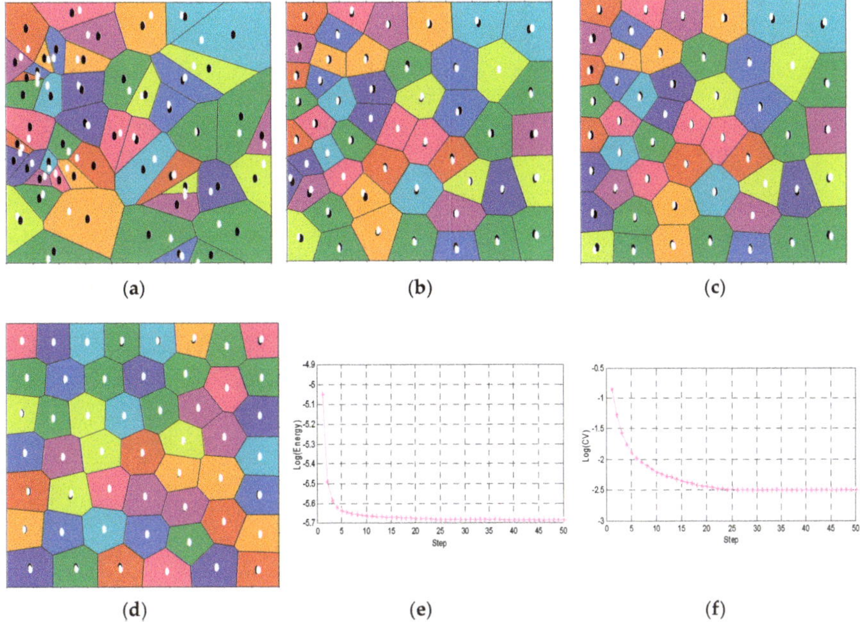

Figure 9. Implementation of the constrained centroidal Voronoi algorithm: (**a**) initial condition, (**b**) step 5, (**c**) step 20, (**d**) step 50, (**e**) change of energy, and (**f**) change of coefficient of variation.

4. Modeling of Columnar Jointed Rock Mass

Analysis of typical shapes of columnar jointing shows that a columnar jointed rock mass can be modelled through the extrusion of a 2-D Voronoi diagram. Hence, the generation of 2-D columnar jointing, which is consistent with the site condition is a key task. In Section 3, it is shown that coefficient of variation can describe the distribution property in the iterations of Lloyd's algorithm. The centroidal Voronoi diagram is applicable in the description of a natural phenomenon such as Giant's Causeway. In fact, natural columnar jointing is not an exactly centroidal Voronoi diagram, and it lies between random and centroidal Voronoi diagrams. In the process of transforming a Voronoi diagram from totally random to centroidal, the value of CV indicates that the Voronoi cells change from heterogeneous to homogeneous. Therefore, it is feasible to describe the characteristics of columnar joints by employing two main parameters: the columnar density representing the scale and coefficient of variation reflecting the variation property.

A procedure for the generation of columnar jointed rock mass is proposed (Figure 10). Based on field investigation of geological conditions and site images, the columnar jointing at the site of interest is characterized by six parameters: columnar density (CD), coefficient of variation (CV), dip direction (DD), dip angle (DIP), transverse joint distance (TD), and probability (TP). The number of random seeds for generating a Voronoi diagram is estimated according to the calculated density. Based on the idea of the bisection method, a novel modified constrained centroidal Voronoi smooth algorithm is implemented to make the CV converge to the specified value as follows:

(1) For a Voronoi diagram with CV larger than the specified value, calculate the centroid P_C.

(2) The new generator $P_{new,g}$ is set at the midpoint of the old generator $P_{old,g}$ and the centroid P_C. Calculate the coefficient of variation CV of the new Voronoi diagram.

(3) Repeat Steps 1 and 2 to obtain the generator until the coefficient of variation CV converges to the specified value with a prescribed accuracy.

Having obtained a 2-D Voronoi diagram with specified CD and CV, it can be extruded to a 3-D columnar shape in accordance with DD and DIP. Then for each extruding Voronoi cell and its length, a transverse joint can be generated by the determination of the intersection between the extruding Voronoi cell and the transverse plane. Sometimes, the transverse joint is non-penetrating, and the transverse joint can be selected with the specified TD and TP. In this way, columnar jointed rock mass with the site properties can be generated.

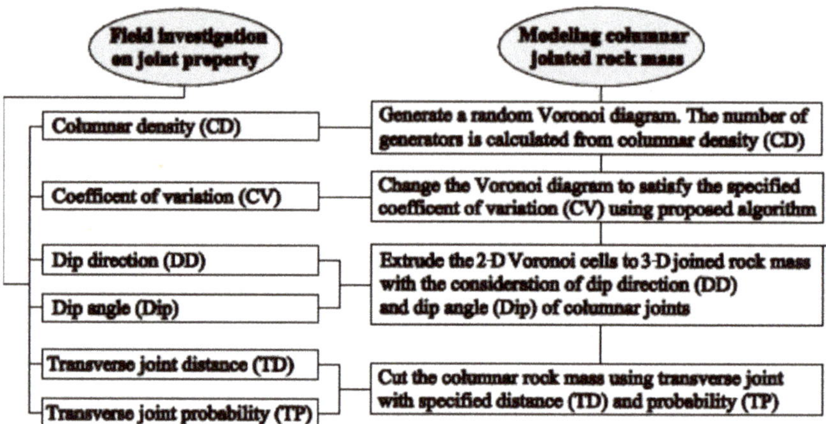

Figure 10. Procedure for modelling columnar joints.

5. Columnar Joints Generation: A Case Study of the Baihetan Hydropower Station

5.1. Engineering Geological Investigation

The Baihetan Hydropower Station is a multi-purpose project for harnessing the Yangtze River, developed mainly for power generation, flood control, sediment flushing, and improving the navigation conditions in the reservoir and downstream. The Baihetan Hydropower Station is located on the downstream reaches of the Jinsha River. At the dam site, the most apparent stratigraphic lithology is basalt that belongs to the Emeishan (Emei Mt.) formation of the Permain System ($P_2\beta$). In the construction of the plant's hydraulic structures, columnar jointed basalt is found to be widely distributed at the arch dam foundation, underground caverns, abutment slopes, and other water conservancy tunnels (Figure 11).

(a)   (b)

Figure 11. Site conditions of the Baihetan Hydropower Station: (**a**) construction site, and (**b**) typical columnar jointed basalt.

A typical figure of columnar jointed basalt is drawn by the Huadong Engineering Cooperation Limited (ECIDI), as shown in Figure 12 with related parameters listed in Table 1. The orientation of columnar is regular with Dip = 72° and DD = 145°, and the length of columnar is about 1.5 m. Based on the skeleton figure, a digital image processing technique is employed in the analysis. The area of each columnar is calculated and the coefficient of variation is obtained. Using a sample of 10 typical columnar basalt drawings, the statistical mean of coefficient of variation was about 44.18% and it implied that the discreteness of columnar basalt was quite large. Furthermore, the density of columnar cells could also be obtained from skeleton drawings, which gave a density of about 20 in a 1 m × 1 m square.

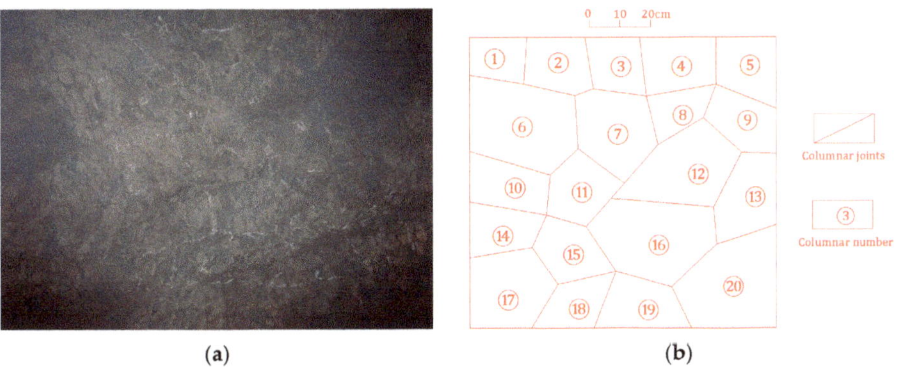

Figure 12. Typical $P_2\beta_3$ columnar basalt at Baihetan Hydropower Station: (**a**) geological photo, and (**b**) joints skeleton.

Table 1. Parameters of columnar jointed rock mass in dam foundation of Baihetan Hydropower Station.

Parameter	Dip	DD	CD	CV	TD	TP
Values	72°	145°	$25/m^2$	44.18%	1.5 m	0.3

5.2. Columnar Jointing Model Generation

Considering the scale effect of heterogeneous materials, a proper size has to be selected. For convenience, the size of the numerical model was taken as 1 m × 1 m. In order to ensure that the 3-D model could be generated, an initial area of 5 m × 5 m was chosen for the generation of a 2-D Voronoi diagram. From the density of site conditions, it was estimated that about 500 seeds were needed. A random Voronoi diagram was generated in Figure 13a with a CV of 56.45%. With a specified CV value of 44.18%, a Voronoi diagram could be generated (Figure 13b) by applying the algorithm proposed in Section 4. Then an extrusion at a certain direction with parameter DIP and DD was made to extend the rock from 2-D to 3-D (Figure 13c). Afterwards, horizontal hidden joints was implemented to cut the rock (Figure 13d). Using these operations, the coordinates information of all rock blocks could be calculated. By trimming the solid with a certain boundary, the block information of the columnar jointed rock is shown in Figure 13e. It was easier for the particle model in Figure 13f, where the model could be generated by importing the DFN (discrete fracture networks) in Figure 13d to a cubic particle model, which could be done easily in DEM packages like PFC (Particle Flow Code), YADE (Yet another dynamic engine), and LIGGGHTS (LAMMPS improved for general granular and granular heat transfer simulations).

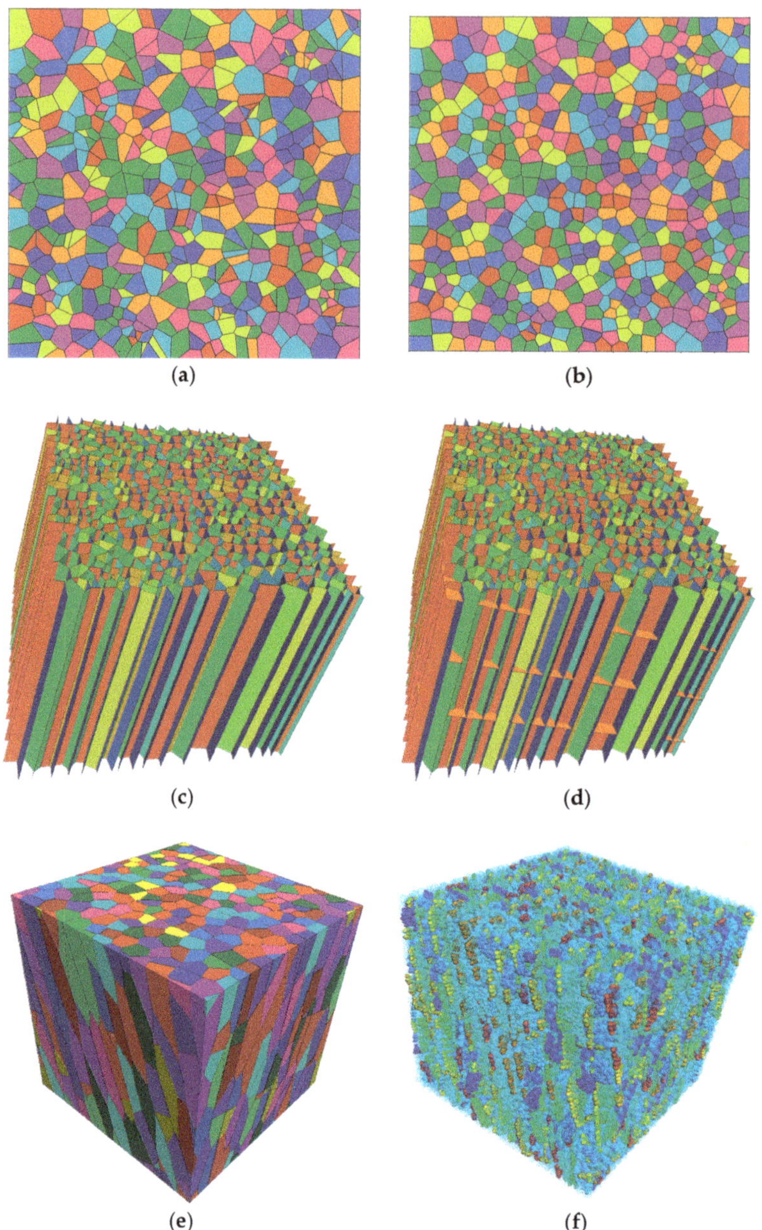

Figure 13. Numerical specimen generation: (**a**) Voronoi diagram with CV = 56.45%, (**b**) Voronoi diagram with CV = 44.18%, (**c**) extrude 2-D Voronoi diagram with direction, (**d**) cut columnar rock with transverse joint, (**e**) block model of columnar rock, and (**f**) particle model of columnar rock.

Columnar jointed rock mass with different CV is shown in Figure 14. It can be seen that the volume of columnar became more and more uniform with the decrease of coefficient of variation. Columnar jointed rock masses with different directions are shown in Figure 15. The result shows that the direction was also an important parameter affecting the morphology of columnar jointed rock masses. From all these cases, it can be concluded that this method is applicable and effective in the modeling of columnar jointed rock mass with complex structures, which is extremely helpful in the analysis of physical and mechanical properties of such materials.

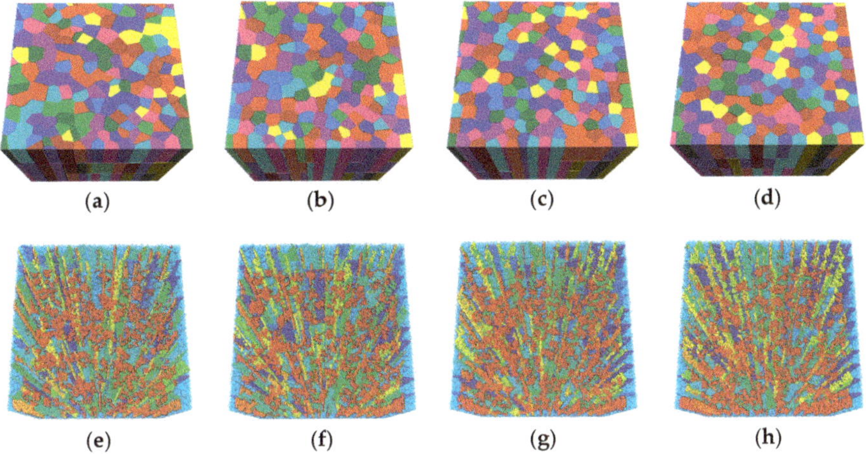

Figure 14. Columnar joints with different coefficient of variation: (**a**) block model with CV = 40%, (**b**) block model with CV = 30%, (**c**) block model with CV – 20%, (**d**) block model with C = 10%, (**e**) particle model with CV = 40%, (**f**) particle model with CV = 30%, (**g**) particle model with CV = 20%, and (**h**) particle model with CV = 10%.

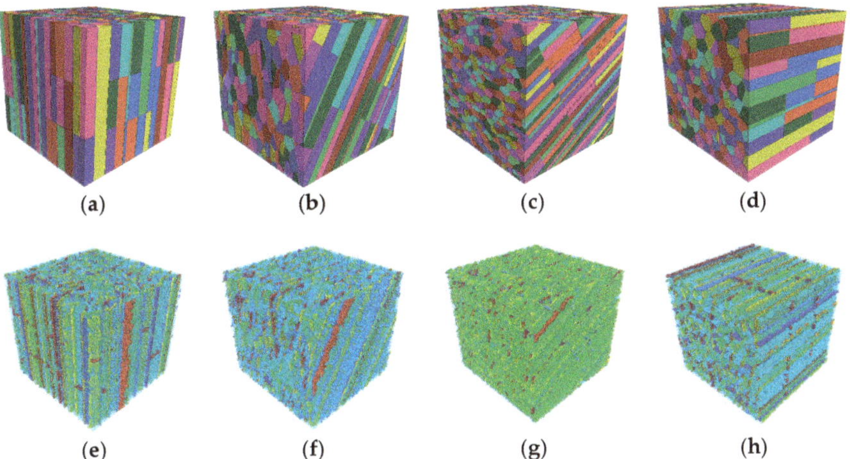

Figure 15. Columnar joints with different joint dip angles: (**a**) block model with dip = 0°, (**b**) block model with dip = 30°, (**c**) block model with dip = 60°, (**d**) block model with dip = 90°, (**e**) particle model with dip = 0°, (**f**) particle model with dip = 30°, (**g**) particle model with dip = 60°, and (**h**) particle model with dip = 90°.

6. Conclusions and Discussion

Aimed at the analysis of columnar jointed rock mass with complex structures, this paper proposed a novel method for the generation of numerical model. From the results obtained, the following conclusions can be drawn:

(1) The coefficient of variation is an effective parameter for representing the deviation between the generator and the centroid of Voronoi cell, which has the same effect as energy in a centroidal Voronoi tessellation. Furthermore, it can reflect the heterogeneity of the cells forming the columnar jointed rock mass.

(2) A modified Lloyd's algorithm is proposed to generate the Voronoi diagram with a specified coefficient of variation. Two algorithms for estimating the centroid were presented and discussed.

(3) This work proposed the description of columnar jointed rock mass with six parameters and a detailed procedure for modelling columnar jointed rock mass. Taking the columnar basalt in the Baihetan hydropower station as an example, numerical models for columnar jointed rock mass with the specified geological properties were generated. The numerical results indicated that the method was effective and efficient.

Author Contributions: Q.M contributed in formulating the research question; Y.C., L.Y. and Q.Z. prepared the original manuscript. All authors contributed in addressing reviewers comments during the review process and proofreading of the accepted manuscript.

Funding: This research was funded by National Key R&D Program of China (2017YFC1501100), the National Natural Science Foundation of China (Grant Nos. 51709089, 11572110, 51479049, 11772116), and Open fund of state key laboratory of hydraulics and mountain river engineering (Sichuan University: SKHL1725) are gratefully acknowledged.

Conflicts of Interest: The authors declare no conflict of interest.

References

1. Goehring, L. *On the Scaling and Ordering of Columnar Joints*; University of Toronto: Toronto, ON, Canada, 2008.
2. Xu, X.W. Columnar jointing in basalt, Shandong Province, China. *J. Struct. Geol.* **2010**, *32*, 1403. [CrossRef]
3. Bulkeley, R. Part of a Letter from Sir RBSRS to Dr. Lister concerning the Giants Causway in the County of Atrim in Ireland. *Philos. Trans. R. Soc. Lond.* **1693**, *17*, 708–710.
4. Budkewitsch, P.; Robin, P.Y. Modelling the evolution of columnar joints. *J. Volcanol. Geotherm. Res.* **1994**, *59*, 219–239. [CrossRef]
5. DeGraff, J.M.; Long, P.E.; Aydin, A. Use of joint-growth directions and rock textures to infer thermal regimes during solidification of basaltic lava flows. *J. Volcanol. Geotherm. Res.* **1989**, *38*, 309–324. [CrossRef]
6. Ryan, M.P.; Sammis, C.G. Cyclic fracture mechanisms in cooling basalt. *Geol. Soc. Am. Bull.* **1978**, *89*, 1295–1308. [CrossRef]
7. Ryan, M.P.; Sammis, C.G. The glass transition in basalt. *J. Geophys. Res. Solid Earth* **1981**, *86*, 9519–9535. [CrossRef]
8. Müller, G. Experimental simulation of basalt columns. *J. Volcanol. Geotherm. Res.* **1998**, *86*, 93–96. [CrossRef]
9. Müller, G. Experimental simulation of joint morphology. *J. Struct. Geol.* **2001**, *23*, 45–49. [CrossRef]
10. Shorlin, K.A.; de Bruyn, J.R.; Graham, M.; Morris, S.W. Development and geometry of isotropic and directional shrinkage-crack patterns. *Phys. Rev. E* **2000**, *61*, 6950. [CrossRef]
11. Jiang, Q.; Feng, X.T.; Hatzor, Y.H.; Hao, X.J.; Li, S.J. Mechanical anisotropy of columnar jointed basalts: An example from the Baihetan hydropower station, China. *Eng. Geol.* **2014**, *175*, 35–45. [CrossRef]
12. King, M.; Myer, L.; Rezowalli, J. Experimental studies of elastic-wave propagation in a columnar-jointed rock mass. *Geophys. Prospect.* **1986**, *34*, 1185–1199. [CrossRef]
13. Meng, G.T. *Geomechanical Analysis of Anisotropic Columnar Jointed Rock Mass and Its Application in Hydropower Engineering*; Hohai University: Nanjing, China, 2007.
14. Zheng, W.T.; Xu, W.Y.; Yan, D.X.; Ji, H. *A Three-Dimensional Modeling Method for Irregular Columnar Joints Based on Voronoi Graphics Theory. Applied Informatics and Communication*; Springer: Berlin/Heidelberg, Germany, 2011; pp. 62–69.
15. Di, S.J.; Xu, W.Y.; Ning, Y.; Wang, W.; Wu, G.Y. Macro-mechanical properties of columnar jointed basaltic rock masses. *J. Cent. South Univ. Technol.* **2011**, *18*, 2143–2149. [CrossRef]
16. Shan, Z.G.; Di, S.J. Loading-unloading test analysis of anisotropic columnar jointed basalts. *J. Zhejiang Univ. Sci. A* **2013**, *14*, 603–614. [CrossRef]

17. Yan, D.X.; Xu, W.Y.; Zheng, W.T.; Wang, W.; Shi, A.C.; Wu, G.Y. Mechanical characteristics of columnar jointed rock at dam base of Baihetan hydropower station. *J. Cent. South Univ. Technol.* **2011**, *18*, 2157–2162. [CrossRef]
18. Xu, W.Y.; Zheng, W.T.; Ning, Y.; Meng, G.T.; Wu, G.Y.; Shi, A.C. 3D anisotropic numerical analysis of rock mass with columnar joints for dam foundation. *Yantu Lixue* **2010**, *31*, 949–955.
19. Xiao, W.M.; Deng, R.G.; Zhong, Z.B.; Fu, X.M.; Wang, C.Y. Experimental study on the mechanical properties of simulated columnar jointed rock masses. *J. Geophys. Eng.* **2015**, *12*, 80. [CrossRef]
20. Ning, Y.; Xu, W.Y.; Zheng, W.T. Study of random simulation of columnar jointed rock mass and its representative elementary volume scale. *Chin. J. Rock Mech. Eng.* **2008**, *27*, 1202–1208.
21. O'Reilly, J.P. Explanatory Notes and Discussion of the Nature of the Prismatic Forms of a Group of Columnar Basalts, Giant's Causeway. (With Plates XV.-XVIII.). *Trans. R. Ir. Acad.* **1879**, *26*, 641–734.
22. Mollon, G.; Zhao, J. Fourier–Voronoi-based generation of realistic samples for discrete modelling of granular materials. *Granul. Matter* **2012**, *14*, 621–638. [CrossRef]
23. Du, Q.; Faber, V.; Gunzburger, M. Centroidal Voronoi Tessellations: Applications and Algorithms. *SIAM Rev.* **1999**, *41*, 637–676. [CrossRef]
24. Du, Q.; Wong, T.W. Numerical studies of MacQueen's k-means algorithm for computing the centroidal Voronoi tessellations. *Comput. Math. Appl.* **2002**, *44*, 511–523. [CrossRef]
25. Lloyd, S.P. Least squares quantization in PCM. *IEEE Trans. Inf. Theory* **1982**, *28*, 129–137. [CrossRef]
26. Barber, C.B.; Dobkin, D.P.; Huhdanpaa, H. The quickhull algorithm for convex hulls. ACM Trans. *Math. Softw.* **1996**, *22*, 469–483. [CrossRef]
27. Duyckaerts, C.; Godefroy, G. Voronoi tessellation to study the numerical density and the spatial distribution of neurones. *J. Chem. Neuroanat.* **2000**, *20*, 83–92. [CrossRef]
28. Minciacchi, D.; Granato, A. How relevant are subcortical maps for the cortical machinery? An hypothesis based on parametric study of extra-relay afferents to primary sensory areas. *Adv. Psychol.* **1997**, *119*, 149–168.
29. Burkardt, J. 2015. Available online: http://people.sc.fsu.edu/~jburkardt/m_src/cvt_2d_sampling/cvt_2d_sampling.html (accessed on 21 October 2018).

Article

Effect of Plastic Anisotropy on the Distribution of Residual Stresses and Strains in Rotating Annular Disks

Woncheol Jeong [1,*], Sergei Alexandrov [2,3] and Lihui Lang [2]

[1] Department of Materials Science and Engineering, Seoul National University, 1 Gwanak-ro, Gwanak-gu, Seoul 08826, Korea

[2] School of Mechanical Engineering and Automation, Beihang University, No. 37 Xueyuan Road, Beijing 100191, China; sergei_alexandrov@spartak.ru (S.A.); lang@buaa.edu.cn (L.L.)

[3] Ishlinskii Institute for Problems in Mechanics, 101-1 Prospect Vernadskogo, 119526 Moscow, Russia

* Correspondence: rippler@snu.ac.kr

Received: 6 September 2018; Accepted: 18 September 2018; Published: 19 September 2018

Abstract: Hill's quadratic orthotropic yield criterion is used for revealing the effect of plastic anisotropy on the distribution of stresses and strains within rotating annular polar orthotropic disks of constant thickness under plane stress. The associated flow rule is adopted for connecting the stresses and strain rates. Assuming that unloading is purely elastic, the distribution of residual stresses and strains is determined as well. The solution for strain rates reduces to one nonlinear ordinary differential equation and two linear ordinary differential equations, even though the boundary value problem involves two independent variables. The aforementioned differential equations can be solved one by one. This significantly simplifies the numerical treatment of the general boundary value problem and increases the accuracy of its solution. In particular, comparison with a finite difference solution is made. It is shown that the finite difference solution is not accurate enough for some applications.

Keywords: plastic anisotropy; rotating disk; plane stress; residual stresses and strains; flow theory of plasticity; semi-analytic solution

---

1. Introduction

Thin rotating disks are used in many applications such as energy storage devices; gyroscopic control devices for ships, submarines, aircrafts, rockets, and missiles; high-speed gears; and turbine rotors [1]. Moreover, rotational autofrettage has been recently proposed [2] as a new technique for producing compressive residual stresses. A purely elastic solution for isotropic rotating disks has been given [3], and a comprehensive overview of the problem of an elastic rotating disk up through to the late 1990s has been provided [4]. Further, an elastic solution for arbitrarily functionally graded polar orthotropic rotating disks has been recently proposed [5].

A great number of elastic/plastic solutions have been also reported in the literature. In most cases it has been assumed that plastic yielding is controlled by the Tresca yield criterion. A review of such solutions has been provided [6]. A few solutions for the deformation theory of plasticity based on the von Mises yield criterion are also available; a review of these solutions has been given [7]. In the case of the flow theory of plasticity, the finite difference method has usually been adopted for determining the distribution of strains [8–10]. An efficient method that advances the analytical treatment of elastic/plastic rotating disks has been proposed [11] for the von Mises yield criterion and its associated flow rule. However, the general idea of the method can be extended to other yield criteria with no difficulty. In particular, it has been demonstrated [12] where a model of pressure-dependent plasticity has been adopted.

Solutions for anisotropic materials are of special importance because even mild plastic anisotropy has an amplified effect on residual stress and strain distributions under certain conditions [13]. The proposed orthotropic yield criterion [14] is often used to model plastic anisotropy in rotating disks, for example [15,16]. An important type of anisotropy is polar anisotropy (see, for example, [16–20]). Therefore, the present paper deals with polar orthotropic disks obeying the yield criterion [14]. The method developed in [11] is employed. This allows a semi-analytic solution to be found. Comparison with the finite difference solution presented in [10] is made. It is shown that the finite difference solution is not accurate enough for some applications. This demonstrates an advantage of using the method [11] as compared with the finite difference method.

2. Statement of the Problem

A detailed description of the boundary value problem under consideration can be found in many works (see, for example, [11]). The boundary value problem is solved in a cylindrical coordinate system (r, θ, z) whose z axis coincides with the axis of symmetry of a thin annular rotating disk of constant thickness. The assumption of constant thickness is often adopted in theoretical analyses of rotating disks [21–28]. The outer and inner radii of the disk are denoted as b_0 and a_0, respectively. The angular velocity of the disk is ω. The boundary value problem is illustrated in Figure 1. Strains are small. The disk has no stress at $\omega = 0$. The normal stresses in the cylindrical coordinate system, σ_r, σ_θ, and σ_z, are the principal stresses. The solution of the boundary value problem is independent of the polar angle. The state of stress is plane, $\sigma_z = 0$. The component of the acceleration vector in the circumferential direction is neglected. The stress boundary conditions are:

$$\sigma_r = 0 \tag{1}$$

for $r = a_0$ and $r = b_0$. In general, the disk consists of two regions: elastic and plastic. Hooke's law connects the elastic strains and stresses. In particular,

$$\varepsilon_r^e = \frac{\sigma_r - \nu\sigma_\theta}{E}, \quad \varepsilon_\theta^e = \frac{\sigma_\theta - \nu\sigma_r}{E}, \quad \varepsilon_z^e = -\frac{\nu(\sigma_r + \sigma_\theta)}{E}. \tag{2}$$

Here, ν is Poisson's ratio and E is Young's modulus, ε_r is the radial strain, ε_θ is the circumferential strain, and ε_z is the axial strain. The superscript e denotes the elastic part of the strain and will denote the elastic part of the strain rate. The orthotropic yield criterion proposed in reference [14] and its associated flow rule are adopted in the plastic region. It is assumed that the principal axes of anisotropy coincide with coordinate curves of the cylindrical coordinate system. Under plane stress conditions, the yield criterion adopted reads $(G + H)\sigma_r^2 + (F + H)\sigma_\theta^2 - 2H\sigma_r\sigma_\theta = 1$ where G, H, and F are anisotropic constants. It is convenient to rewrite this criterion as:

$$\sigma_r^2 + \frac{\sigma_\theta^2}{\eta_1^2} - \frac{\eta\sigma_r\sigma_\theta}{\eta_1} = \sigma_0^2 \tag{3}$$

where

$$\eta = \frac{2H}{\sqrt{(G+H)(H+F)}}, \quad \eta_1 = \frac{\sqrt{G+H}}{\sqrt{H+F}}, \quad \eta_2 = \frac{2F}{(H+F)\sqrt{4-\eta^2}}, \quad \sigma_0 = \frac{1}{G+H}. \tag{4}$$

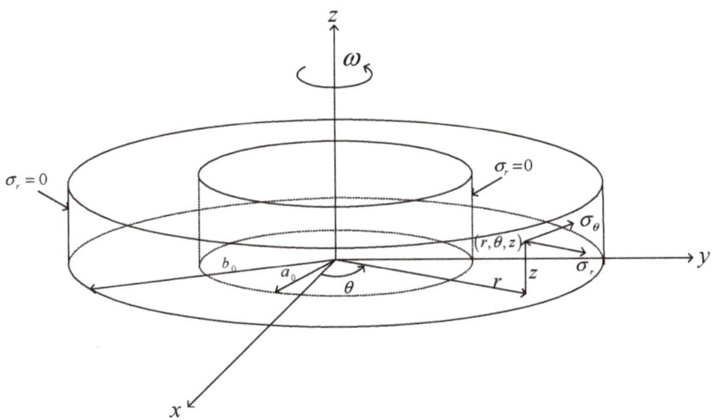

Figure 1. Schematic diagram of a rotating annular disk.

The yield criterion (Equation (3)) reduces to the von Mises yield criterion at $F = G = H$. In this case, σ_0 is the tensile yield stress. Let $\dot{\varepsilon}_r^p$, $\dot{\varepsilon}_\theta^p$, and $\dot{\varepsilon}_z^p$ be the plastic strain rates. In the case under consideration, the associated flow rule can be written as [29]:

$$\dot{\varepsilon}_r^p = \lambda\left(\sigma_r - \frac{\eta}{2\eta_1}\sigma_\theta\right), \quad \dot{\varepsilon}_\theta^p = \frac{\lambda}{\eta_1}\left(\frac{\sigma_\theta}{\eta_1} - \frac{\eta}{2}\sigma_r\right), \quad \dot{\varepsilon}_z^p = -\lambda\left[\left(1 - \frac{\eta}{2\eta_1}\right)\sigma_r + \left(\frac{1}{\eta_1} - \frac{\eta}{2}\right)\frac{\sigma_\theta}{\eta_1}\right] \tag{5}$$

where λ is a non-negative multiplier. The superimposed dot denotes the time derivative at fixed r and the superscript p denotes the plastic part of the strain rate and will denote the plastic part of the strain. The total strains and strain rates in the plastic region are:

$$\begin{aligned}
\varepsilon_r &= \varepsilon_r^e + \varepsilon_r^p, \quad \varepsilon_\theta = \varepsilon_\theta^e + \varepsilon_\theta^p, \quad \varepsilon_z = \varepsilon_z^e + \varepsilon_z^p, \\
\dot{\varepsilon}_r &= \dot{\varepsilon}_r^e + \dot{\varepsilon}_r^p, \quad \dot{\varepsilon}_\theta = \dot{\varepsilon}_\theta^e + \dot{\varepsilon}_\theta^p, \quad \dot{\varepsilon}_z = \dot{\varepsilon}_z^e + \dot{\varepsilon}_z^p.
\end{aligned} \tag{6}$$

The equilibrium equation is of the form:

$$\frac{\partial\sigma_r}{\partial r} + \frac{\sigma_r - \sigma_\theta}{r} = -\varsigma\omega^2 r. \tag{7}$$

Here, ς is the mass density of the material.

The boundary value problem is classified by the following dimensionless quantities:

$$\rho = \frac{r}{b_0}, \quad \Omega = \frac{\varsigma\omega^2 b_0^2}{\sigma_0}, \quad a = \frac{a_0}{b_0}, \quad k = \frac{\sigma_0}{E}. \tag{8}$$

Since the material model is rate-independent, the time derivative can be replaced with the derivative with respect to Ω. The derivatives of strain components with respect to Ω are denoted as:

$$\begin{aligned}
\xi_r &= \frac{\partial\varepsilon_r}{\partial\Omega}, \quad \xi_\theta = \frac{\partial\varepsilon_\theta}{\partial\Omega}, \quad \xi_z = \frac{\partial\varepsilon_z}{\partial\Omega}, \\
\xi_r^e &= \frac{\partial\varepsilon_r^e}{\partial\Omega}, \quad \xi_\theta^e = \frac{\partial\varepsilon_\theta^e}{\partial\Omega}, \quad \xi_z^e = \frac{\partial\varepsilon_z^e}{\partial\Omega}, \\
\xi_r^p &= \frac{\partial\varepsilon_r^p}{\partial\Omega}, \quad \xi_\theta^p = \frac{\partial\varepsilon_\theta^p}{\partial\Omega}, \quad \xi_z^p = \frac{\partial\varepsilon_z^p}{\partial\Omega}.
\end{aligned} \tag{9}$$

Then, the equation of strain rate compatibility is equivalent to:

$$\rho\frac{\partial\xi_\theta}{\partial\rho} = \xi_r - \xi_\theta. \tag{10}$$

Using Equation (8), Equation (7) can be rewritten as:

$$\frac{\partial \sigma_r}{\sigma_0 \partial \rho} + \frac{\sigma_r - \sigma_\theta}{\sigma_0 \rho} = -\Omega \rho. \tag{11}$$

3. Solution at Loading

3.1. Purely Elastic Solution

The general purely elastic solution for a rotating disk is given, for example, in [3]. Using Equation (8), the solution satisfying the boundary condition (Equation (1)) at $\rho = 1$ is represented as:

$$
\begin{aligned}
\frac{\sigma_r}{\sigma_0} &= B\left(\frac{1}{\rho^2} - 1\right) + \frac{\Omega(3+\nu)}{8}\left(1 - \rho^2\right), \\
\frac{\sigma_\theta}{\sigma_0} &= -B\left(\frac{1}{\rho^2} + 1\right) + \frac{\Omega(1+3\nu)}{8}\left(\frac{3+\nu}{1+3\nu} - \rho^2\right), \\
\frac{\varepsilon_r}{k} &= \frac{8B\left[1+\nu-(1-\nu)\rho^2\right]+\Omega(1-\nu)\left[3+\nu-3(1+\nu)\rho^2\right]\rho^2}{8\rho^2}, \\
\frac{\varepsilon_\theta}{k} &= \frac{-8B\left[1+\nu+(1-\nu)\rho^2\right]+\Omega(1-\nu)\left[3+\nu-(1+\nu)\rho^2\right]\rho^2}{8\rho^2}, \\
\frac{\varepsilon_z}{k} &= \frac{\nu}{4}\left\{8B - \Omega\left[3+\nu-2(1+\nu)\rho^2\right]\right\}.
\end{aligned}
\tag{12}
$$

Here, B is a constant of integration. Using the boundary condition (Equation (1)) at $\rho = a$, it is possible to find that:

$$B = -\frac{\Omega(3+\nu)a^2}{8}. \tag{13}$$

Eliminating B in Equation (12) by means of Equation (13) provides the solution for the stresses and strains in the purely elastic disk. Let Ω_e be the value of Ω at which the plastic region starts to develop from the surface $\rho = a$. Substituting the values of σ_r and σ_θ at $\rho = a$ into the yield condition Equation (3) and using Equation (4) yields:

$$\Omega_e = \frac{4\eta_1}{3 + \nu + a^2(1 - \nu)}. \tag{14}$$

In what follows, it is assumed that $\Omega > \Omega_e$.

The solution (Equation (12)) is valid in the elastic region of the elastic/plastic disk. However, Equation (13) is not valid in this case because one of the boundary conditions in Equation (1) should be replaced with the condition that the radial and circumferential stresses are continuous across the elastic/plastic boundary.

3.2. Elastic/Plastic Stress Solution

The elastic/plastic stress solution is available [30]. For completeness, this solution is outlined below. The yield condition (Equation (3)) is satisfied by the following substitution:

$$\sigma_r/\sigma_0 = 2\cos\psi/\sqrt{4 - \eta^2}, \quad \sigma_\theta/\sigma_0 = \left(\eta\cos\psi/\sqrt{4 - \eta^2} + \sin\psi\right)\eta_1 \tag{15}$$

where ψ is a new function of ρ and Ω. Substituting Equation (15) into Equation (11) and using Equation (4) leads to:

$$\frac{2\sin\psi}{\sqrt{4 - \eta^2}}\frac{\partial\psi}{\partial\rho} + \frac{(\eta_2\cos\psi - \eta_1\sin\psi)}{\rho} = \Omega\rho. \tag{16}$$

The boundary condition to this equation is determined from the boundary condition (Equation (1)) at $\rho = a$ and Equation (15) as:

$$\psi = \frac{\pi}{2} \tag{17}$$

for $\rho = a$. It has been taken into account here that $\sigma_\theta > 0$ at $\rho = a$. Let ρ_c be the dimensionless radius of the elastic/plastic boundary. The value of ψ at $\rho = \rho_c$ is denoted by ψ_c. The continuity of the radial and circumferential stresses across the elastic/plastic boundary, together with Equations (12) and (15), leads to:

$$B\left(\frac{1}{\rho_c^2} - 1\right) + \frac{\Omega(3+v)}{8}\left(1 - \rho_c^2\right) = 2\cos\psi_c/\sqrt{4-\eta^2},$$

$$-B\left(\frac{1}{\rho_c^2} + 1\right) + \frac{\Omega(1+3v)}{8}\left(\frac{3+v}{1+3v} - \rho_c^2\right) = \left(\eta\cos\psi_c/\sqrt{4-\eta^2} + \sin\psi_c\right)\eta_1. \tag{18}$$

Eliminating B between these equations yields:

$$\frac{2\cos\psi_c}{\sqrt{4-\eta^2}} + \frac{\left(\eta\cos\psi_c/\sqrt{4-\eta^2}+\sin\psi_c\right)\left(1-\rho_c^2\right)\eta_1}{\left(1+\rho_c^2\right)}$$
$$-\frac{\Omega\left(1-\rho_c^2\right)}{8}\left[\frac{(1+3v)}{\left(1+\rho_c^2\right)}\left(\frac{3+v}{1+3v} - \rho_c^2\right) + 3 + v\right] = 0. \tag{19}$$

For a given value of Ω, Equation (19) and the solution of Equation (16) supply the system of equations for ρ_c and ψ_c. Then, B can be determined from Equation (18). The stress distribution in the elastic region, $\rho_c \leq \rho \leq 1$, follows from Equation (12). The stress distribution in the plastic region, $a \leq \rho \leq \rho_c$, is readily determined from Equation (15) and the solution of Equation (16). The latter is in parametric form with ψ being the parameter. The plastic region occupies the entire disk when $\rho_c = 1$. Let Ω_p be the corresponding value of Ω. This value can be found numerically using the dependence of ρ_c on Ω known from Equation (19) and the solution of Equation (16). In particular, it is seen from Equation (19) that $\psi_c = \pi/2$ at $\rho_c = 1$. This condition and Equation (17) allow the value of Ω_p and a constant of integration to be determined from the solution of Equation (16).

3.3. Elastic/Plastic Strain Solution

Eliminating B in Equation (12) by means of the solution of Equations (18) and (19) supplies the strain solution in the elastic region as follows in terms of Ω and ρ. Replacing the plastic strain rates in Equation (5) with the corresponding quantities introduced in Equation (9) and eliminating λ between the resulting equations leads to:

$$\xi_r^p = \xi_\theta^p\frac{(2\eta_1\sigma_r - \eta\sigma_\theta)\eta_1}{(2\sigma_\theta - \eta\eta_1\sigma_r)}, \qquad \xi_z^p = \xi_\theta^p\frac{[(2\eta_1 - \eta)\eta_1\sigma_r + (2 - \eta\eta_1)\sigma_\theta]}{(\eta\eta_1\sigma_r - 2\sigma_\theta)}.$$

The stresses in these equations can be expressed in terms of ψ by means of Equation (15). Then,

$$\xi_r^p = \xi_\theta^p\frac{\left(\sqrt{4-\eta^2}\cos\psi - \eta\sin\psi\right)\eta_1}{2\sin\psi}, \qquad \xi_z^p = -\xi_\theta^p\frac{\left[\eta_1\sqrt{4-\eta^2}\cos\psi + (2-\eta\eta_1)\sin\psi\right]}{2\sin\psi}. \tag{20}$$

The elastic portion of the strain components in the plastic region is determined from Equations (2), (8) and (15) as:

$$\frac{\varepsilon_r^e}{k} = \frac{(2-\eta\eta_1 v)\cos\psi}{\sqrt{4-\eta^2}} - \eta_1 v\sin\psi, \qquad \frac{\varepsilon_\theta^e}{k} = \frac{(\eta\eta_1 - 2v)\cos\psi}{\sqrt{4-\eta^2}} + \eta_1\sin\psi,$$

$$\frac{\varepsilon_z^e}{k} = \frac{v(\eta\eta_1 + 2)\cos\psi}{\sqrt{4-\eta^2}} + v\eta_1\sin\psi. \tag{21}$$

Differentiating these expressions with respect to Ω and using Equation (9) yields:

$$\frac{\xi_r^e}{k} = -\left[\frac{(2-\eta\eta_1 v)\sin\psi}{\sqrt{4-\eta^2}} + \eta_1 v\cos\psi\right]\frac{\partial\psi}{\partial\Omega}, \qquad \frac{\xi_\theta^e}{k} = \left[\eta_1\cos\psi - \frac{(\eta\eta_1 - 2v)\sin\psi}{\sqrt{4-\eta^2}}\right]\frac{\partial\psi}{\partial\Omega},$$

$$\frac{\xi_z^e}{k} = v\left[\eta_1\cos\psi - \frac{(\eta\eta_1 + 2)\sin\psi}{\sqrt{4-\eta^2}}\right]\frac{\partial\psi}{\partial\Omega}. \tag{22}$$

Taking into account Equation (6), Equation (10) can be rewritten as $\rho\partial\zeta_\theta/\partial\rho = \zeta_r^p + \zeta_r^e - \zeta_\theta^p - \zeta_\theta^e$. Eliminating ζ_r^p in this equation by means of Equation (20) results in:

$$\rho\frac{\partial\zeta_\theta}{\partial\rho} = \frac{\left[\eta_1\sqrt{4-\eta^2}\cos\psi - (\eta\eta_1+2)\sin\psi\right]}{2\sin\psi}\zeta_\theta^p + \zeta_r^e - \zeta_\theta^e \ .$$

This equation and Equation (6) combine to give the following equation for ζ_θ:

$$\rho\frac{\partial\zeta_\theta}{\partial\rho} = \frac{\left[\eta_1\sqrt{4-\eta^2}\cos\psi - (\eta\eta_1+2)\sin\psi\right]}{2\sin\psi}\zeta_\theta + \frac{\eta_1\left[\eta\sin\psi - \sqrt{4-\eta^2}\cos\psi\right]}{2\sin\psi}\zeta_r^e - \zeta_\theta^e \tag{23}$$

where ζ_r^e and ζ_θ^e should be eliminated by means of Equation (22). It is therefore evident that Equation (23) is a linear ordinary differential equation. In order to solve this equation, it is necessary to express the derivative $\partial\psi/\partial\Omega$ involved in Equation (22) in terms of ψ or/and ρ. Following the method proposed in [8], Equation (16) is differentiated with respect to Ω. Then,

$$\frac{2\sin\psi}{\sqrt{4-\eta^2}}\frac{\partial\chi}{\partial\rho} + \left(\frac{2\cos\psi}{\sqrt{4-\eta^2}}\frac{\partial\psi}{\partial\rho} + \frac{\eta_2\sin\psi + \eta_1\cos\psi}{\rho}\right)\chi - \rho = 0 \tag{24}$$

where $\chi = \partial\psi/\partial\Omega$. Using Equation (16), the derivative $\partial\psi/\partial\rho$ involved in Equation (24) can be expressed in terms of ψ and ρ. Then, Equation (24) becomes:

$$\frac{2\sin\psi}{\sqrt{4-\eta^2}}\frac{\partial\chi}{\partial\rho} + \left\{\frac{\eta_2\sin\psi + \eta_1\cos\psi}{\rho} + 2\cos\psi\left[\Omega\rho - \frac{(\eta_2\cos\psi - \eta_1\sin\psi)}{2\rho\sin\psi}\right]\right\}\chi - \rho = 0. \tag{25}$$

It follows from the boundary condition (Equation (17)) that $\partial\psi/\partial\Omega = 0$ at $\rho = a$. Therefore,

$$\chi = 0 \tag{26}$$

at $\rho = a$. This is the boundary condition to Equation (25), which is a linear ordinary differential equation for χ. This equation should be solved numerically. Once χ has been found from Equation (25), it is possible to determine ζ_r^e and ζ_θ^e involved in Equation (23) as functions of ψ and ρ by means of Equation (22). Having ψ as a function of ρ due to the solution of Equation (16), it is possible to represent the coefficients of Equation (23) as functions of ρ. Then, this ordinary differential equation can be solved numerically with no difficulty. The boundary condition to Equation (23) follows from the continuity of the circumferential strain rate and, therefore, ζ_θ across the elastic/plastic boundary $\rho = \rho_c$. The value of ζ_θ on the elastic side of this boundary is determined from Equation (12) as:

$$\frac{\zeta_c}{k} = \frac{(1-v)\left[3+v-(1+v)\rho_c^2\right]}{8} - \frac{\left[1+v+(1-v)\rho_c^2\right]}{\rho_c^2}\frac{dB}{d\Omega}. \tag{27}$$

Then, the boundary condition to Equation (23) is

$$\zeta_\theta = \zeta_c \tag{28}$$

for $\rho = \rho_c$. Once the solution of Equation (23) has been found, the total circumferential strain in the plastic region is determined by integration of ζ_θ with respect to Ω at a given point $\rho = \rho_t$. The maximum value of Ω is denoted as Ω_f and the corresponding value of ρ_c as ρ_f. It is assumed that $\Omega_f < \Omega_p$. It is obvious that $a \le \rho_t < \rho_f$. The value of Ω and the value of the circumferential strain on the elastic side of the elastic/plastic boundary at $\rho_c = \rho_t$ are denoted as Ω_t and E_θ^t, respectively. Then, it follows from the definition for ζ_θ given in Equation (9) that:

$$\varepsilon_\theta = \int_{\Omega_t}^{\Omega_f} \zeta_\theta d\Omega + E_\theta^t \, . \tag{29}$$

The value of E_θ^t is determined from Equation (12) as:

$$\frac{E_\theta^t}{k} = \frac{-8B_t\left[1 + v + (1-v)\rho_t^2\right] + \Omega_t(1-v)\left[3 + v - (1+v)\rho_t^2\right]\rho_t^2}{8\rho_t^2} \, . \tag{30}$$

Here B_t is the value of B at $\rho_c = \rho_t$. This value follows from Equations (18) and (19). Equations (6) and (22) combine to give:

$$\zeta_\theta^p = \zeta_\theta - k\left[\eta_1 \cos\psi - \frac{(\eta\eta_1 - 2v)\sin\psi}{\sqrt{4 - \eta^2}}\right]\chi \, . \tag{31}$$

Substituting Equation (31) into Equation (20) leads to:

$$
\begin{aligned}
\frac{\zeta_r^p}{k} &= \frac{\eta_1}{2}\left\{\frac{\zeta_\theta}{k} - \left[\eta_1\cos\psi - \frac{(\eta\eta_1 - 2v)\sin\psi}{\sqrt{4-\eta^2}}\right]\chi\right\}\frac{\left(\sqrt{4-\eta^2}\cos\psi - \eta\sin\psi\right)}{\sin\psi}\, , \\[2mm]
\frac{\zeta_z^p}{k} &= -\frac{1}{2}\left\{\frac{\zeta_\theta}{k} - \left[\eta_1\cos\psi - \frac{(\eta\eta_1 - 2v)\sin\psi}{\sqrt{4-\eta^2}}\right]\chi\right\}\frac{\left[\eta_1\sqrt{4-\eta^2}\cos\psi + (2-\eta\eta_1)\sin\psi\right]}{\sin\psi}\, .
\end{aligned}
\tag{32}
$$

Hence,

$$\frac{\varepsilon_r^p}{k} = \int_{\Omega_t}^{\Omega_f}\frac{\zeta_r^p}{k}d\Omega, \quad \frac{\varepsilon_z^p}{k} = \int_{\Omega_t}^{\Omega_f}\frac{\zeta_z^p}{k}d\Omega \, . \tag{33}$$

Here, the integrands are known functions of Ω. In particular, ζ_r^p and ζ_z^p are first eliminated by means of Equation (32). Then, the solutions of Equations (16), (23) and (25) are used to represent ψ, ζ_θ, and χ, respectively, as functions of Ω at any value of $\rho = \rho_t$. Therefore, the integrals involved in Equation (33) can be evaluated numerically. The total radial and axial strains in the plastic zone are found by means of Equation (6) where the plastic parts are given by Equation (33) and the elastic parts by Equation (21).

4. Distribution of Residual Stresses and Strains

It is assumed that unloading is purely elastic (i.e., the distribution of residual stresses found by means of Hooke's law for the increments of stresses does not violate the yield criterion in the range $a \leq \rho \leq 1$). This assumption should be verified a posteriori. The residual stresses are determined as:

$$\sigma_r^{res} = \sigma_r^f + \Delta\sigma_r, \quad \sigma_\theta^{res} = \sigma_\theta^f + \Delta\sigma_\theta \, . \tag{34}$$

Here, σ_r^f and σ_θ^f are the radial and circumferential stresses, respectively, at the end of loading. These stresses were found in the previous section. $\Delta\sigma_r$ and $\Delta\sigma_\theta$ are the increments of the radial and circumferential stresses, respectively, in the course of the process of unloading. Analogously, the residual strains are determined as:

$$\varepsilon_r^{res} = \varepsilon_r^f + \Delta\varepsilon_r, \quad \varepsilon_\theta^{res} = \varepsilon_\theta^f + \Delta\varepsilon_\theta, \quad \varepsilon_z^{res} = \varepsilon_z^f + \Delta\varepsilon_z. \tag{35}$$

Here, ε_r^f, ε_θ^f, and ε_z^f are the total radial, circumferential, and axial strains, respectively, at the end of loading. These strains were found in the previous section. $\Delta\varepsilon_r$, $\Delta\varepsilon_\theta$, and $\Delta\varepsilon_z$ are the increments of the radial, circumferential, and axial strains, respectively, in the course of the process of unloading.

Since the process of unloading is assumed to be purely elastic, the increments of the strains are related by Hooke's law to the increments of the stresses:

$$\Delta\varepsilon_r^e = \frac{\Delta\sigma_r - \nu\Delta\sigma_\theta}{E}, \quad \Delta\varepsilon_\theta^e = \frac{\Delta\sigma_\theta - \nu\Delta\sigma_r}{E}, \quad \Delta\varepsilon_z^e = -\frac{\nu(\Delta\sigma_r + \Delta\sigma_\theta)}{E}. \tag{36}$$

Therefore, the solution (Equations (12) and (13)) in which Ω should be replaced with $-\Omega_f$ is valid. As a result,

$$
\begin{aligned}
\frac{\Delta\sigma_r}{\sigma_0} &= B_f\left(\frac{1}{\rho^2} - 1\right) - \frac{\Omega_f(3+\nu)}{8}\left(1 - \rho^2\right), \\
\frac{\Delta\sigma_\theta}{\sigma_0} &= -B_f\left(\frac{1}{\rho^2} + 1\right) - \frac{\Omega_f(1+3\nu)}{8}\left(\frac{3+\nu}{1+3\nu} - \rho^2\right), \\
\frac{\Delta\varepsilon_r}{k} &= \frac{8B_f\left[1+\nu-(1-\nu)\rho^2\right] - \Omega_f(1-\nu)\left[3+\nu-3(1+\nu)\rho^2\right]\rho^2}{8\rho^2}, \\
\frac{\Delta\varepsilon_\theta}{k} &= \frac{-8B_f\left[1+\nu+(1-\nu)\rho^2\right] - \Omega_f(1-\nu)\left[3+\nu-(1+\nu)\rho^2\right]\rho^2}{8\rho^2}, \\
\frac{\Delta\varepsilon_z}{k} &= \frac{\nu}{4}\left\{8B_f + \Omega_f\left[3+\nu-2(1+\nu)\rho^2\right]\right\}
\end{aligned}
\tag{37}
$$

where

$$B_f = \frac{\Omega_f(3+\nu)a^2}{8}. \tag{38}$$

Substituting the solution found in the previous section together with Equations (37) and (38) into Equation (35) yields the distribution of residual stresses and strains. It follows from Equation (3) that the yield criterion is not violated after unloading if:

$$\left(\frac{\sigma_r^{res}}{\sigma_0}\right)^2 + \left(\frac{\sigma_\theta^{res}}{\sigma_0\eta_1}\right)^2 - \frac{\eta}{\eta_1}\left(\frac{\sigma_r^{res}}{\sigma_0}\right)\left(\frac{\sigma_\theta^{res}}{\sigma_0}\right) - 1 \le 0. \tag{39}$$

Since the distribution of the residual stresses has been found, this inequality can be verified with no difficulty.

5. Illustrative Example

Equations (16), (23) and (24) were solved numerically for several materials whose anisotropic coefficients, given in [31,32], are shown in Table 1. It is assumed that $\nu = 0.3$ in all cases. The value of k introduced in Equation (8) is immaterial. In particular, assume that the solution for a disk of given geometry and physical properties has been found. Then, simple scaling of this solution supplies the solutions for similar disks of material with the same Poisson's ratio and anisotropic coefficients but any value of k. The numerical solution is illustrated for an $a = 0.5$ disk at $\Omega_f = 1.7$. The distribution of stresses is depicted in Figures 2 and 3 and residual stresses in Figures 4 and 5. The associated total and plastic strain fields are shown in Figures 6–11. The variation of the residual strains with ρ is depicted in Figures 12–14. The solution for the residual stresses was used in conjunction with Equation (39) to verify that the process of unloading is purely elastic. It is evident from Figure 3 that the effect of plastic anisotropy on the radius of the elastic plastic boundary is very significant and, as a result, so is the effect of plastic anisotropy on the distribution of stresses, strains, residual stresses, and residual strains. On the other hand, the yield loci for the anisotropic parameters considered are depicted in Figure 15. It is seen from this figure that the yield loci for the materials considered are not very different (except for the yield locus for AA3104). This sensitivity of solutions to the yield locus requires very accurate numerical methods for calculating stress and strain fields. Therefore, it is of interest to compare the present solution and a finite difference solution. The present solution involves fewer approximations than finite difference solutions because the derivative $\partial\psi/\partial\Omega$ is found from Equation (24) without any discretization with respect to Ω (i.e., with respect to time). Therefore, it is natural to assume that the present solution is more accurate. In [10], several finite difference solutions were found for an $a = 3/7$ disk assuming that $\rho_f = 5/7$ (in our nomenclature). Using the anisotropic constants adopted in [10], the distribution of the circumferential strain was determined using the method proposed in the present paper. A comparison of the magnitude of circumferential strain at $\rho = a$ and $\rho_f = 5/7$ (in our nomenclature) predicted by the new method and that found in [10] is presented in Table 2. It can be seen from this table that the accuracy of the finite difference solution may be insufficient for practical applications. In particular, Δ shown in Table 2 is defined as:

$$\Delta = \frac{|\varepsilon_{\theta,FDM} - \varepsilon_{\theta,N}|}{\varepsilon_{\theta,N}} \times 100\%$$

where $\varepsilon_{\theta,FDM}$ is the total circumferential strain at $\rho = a$ and $\rho_f = 5/7$ found in [10] and $\varepsilon_{\theta,N}$ is the total circumferential strain at $\rho = a$ and $\rho_f = 5/7$ found in the present paper.

Table 1. Anisotropic coefficients of several materials.

Material	F/(G + H)	H/(G + H)
DC06	0.243	0.703
AA6016	0.587	0.410
AA5182	0.498	0.419
AA3014	0.239	0.301
Isotropic	0.5	0.5

Table 2. Comparison of the total circumferential strain at $\rho = a$ and $\rho_f = 5/7$ found by the two methods.

F/(G + H)	H/(G + H)	$\varepsilon_{\theta,FDM}$	$\varepsilon_{\theta,N}$	Δ (%)
0.452	0.681	0.00088	0.00134	26.3
0.421	0.615	0.0014	0.0019	34.4
0.283	0.634	0.00178	0.00212	16.0
0.811	0.454	0.0025	0.0036	30.6
0.5	0.5	0.0019	0.0022	13.6

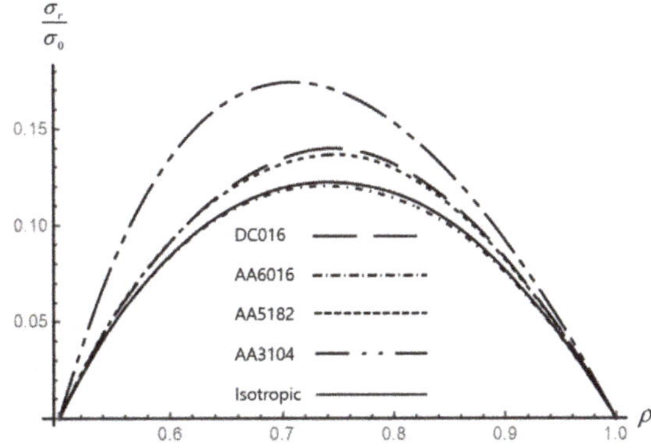

Figure 2. Effect of anisotropic properties on the distribution of the radial stress in an $a = 0.5$ disk.

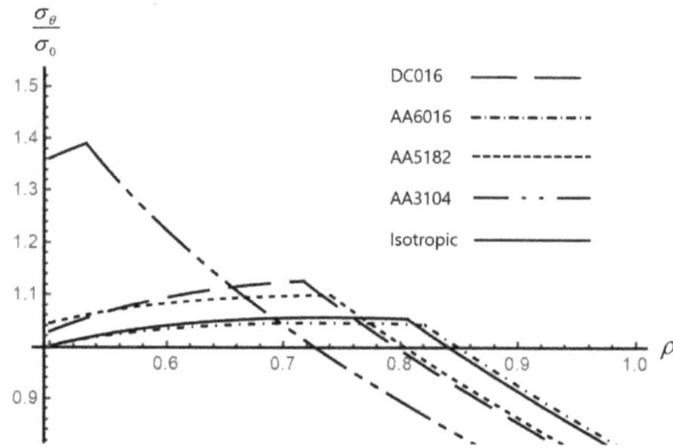

Figure 3. Effect of anisotropic properties on the distribution of the circumferential stress in an $a = 0.5$ disk.

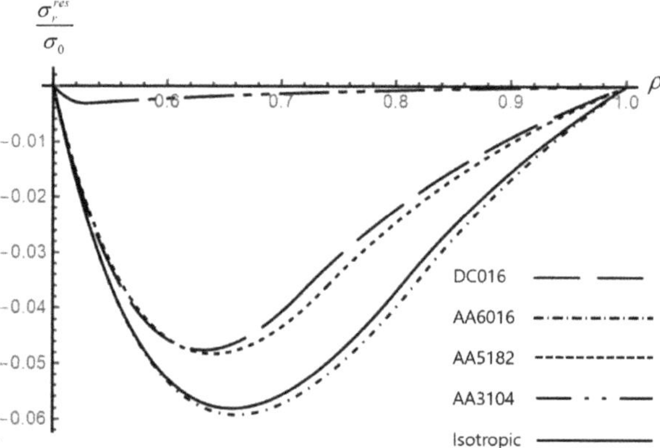

Figure 4. Effect of anisotropic properties on the distribution of the residual radial stress in an $a = 0.5$ disk.

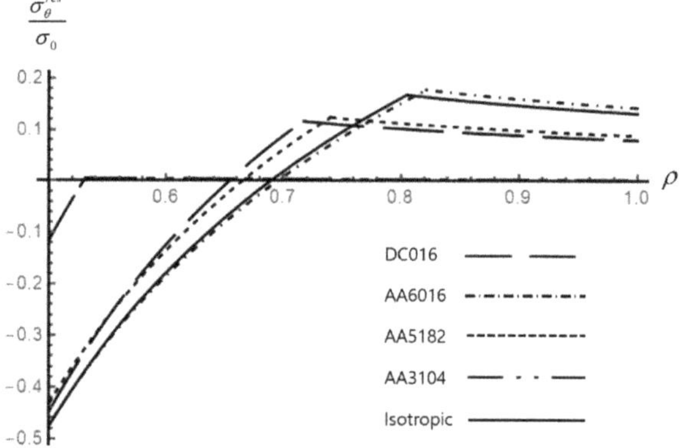

Figure 5. Effect of anisotropic properties on the distribution of the residual circumferential stress in an $a = 0.5$ disk.

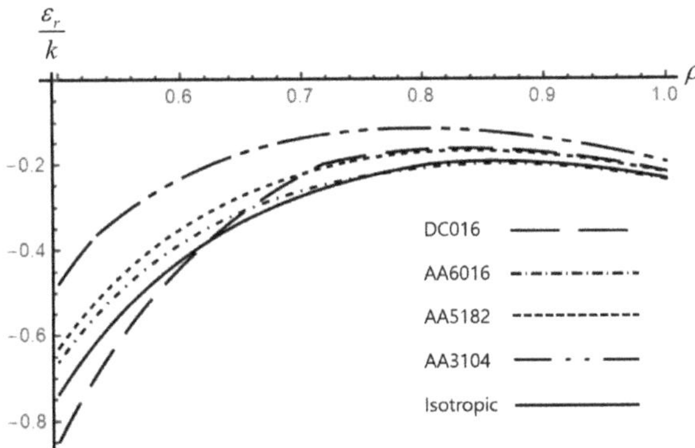

Figure 6. Effect of anisotropic properties on the distribution of the radial strain in an $a = 0.5$ disk.

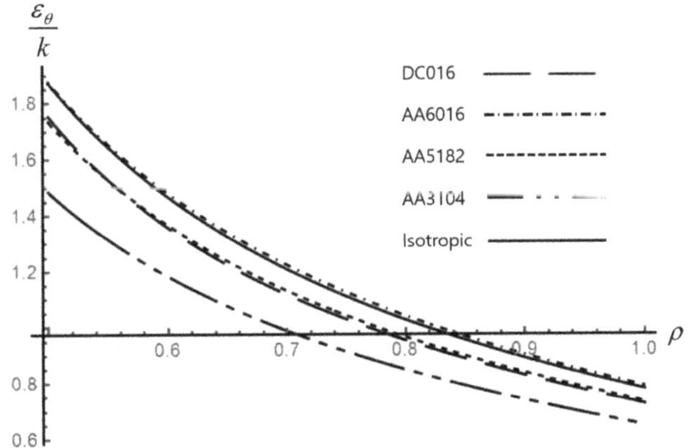

Figure 7. Effect of anisotropic properties on the distribution of the circumferential strain in an $a = 0.5$ disk.

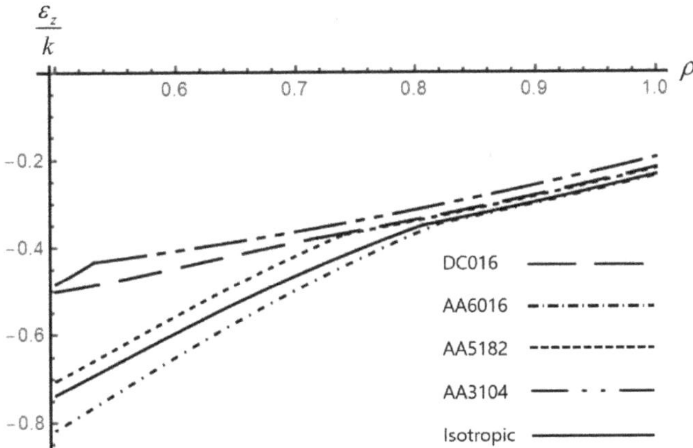

Figure 8. Effect of anisotropic properties on the distribution of the axial strain in an $a = 0.5$ disk.

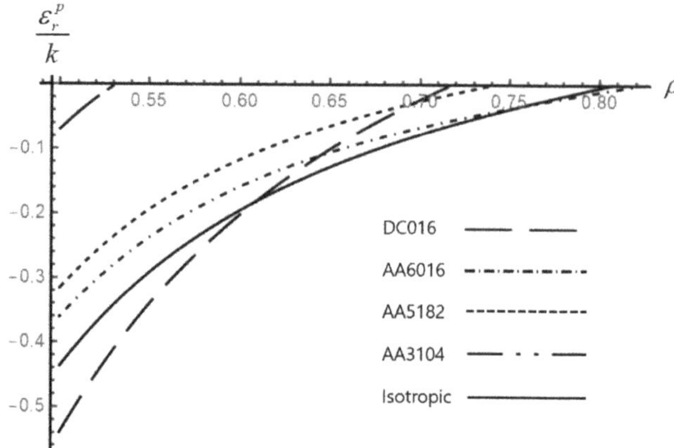

Figure 9. Effect of anisotropic properties on the distribution of the radial plastic strain in an $a = 0.5$ disk.

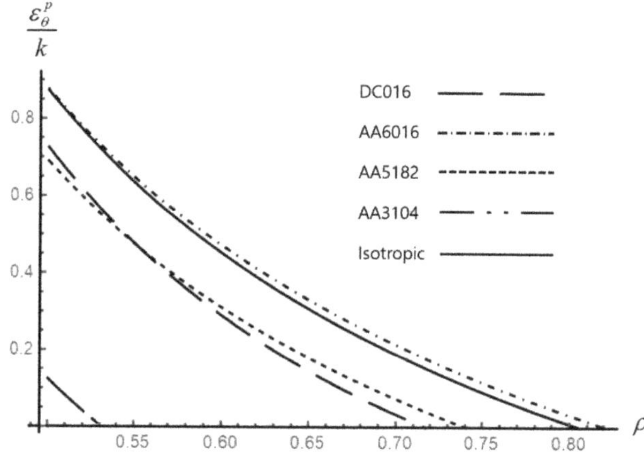

Figure 10. Effect of anisotropic properties on the distribution of the circumferential plastic strain in an $a = 0.5$ disk.

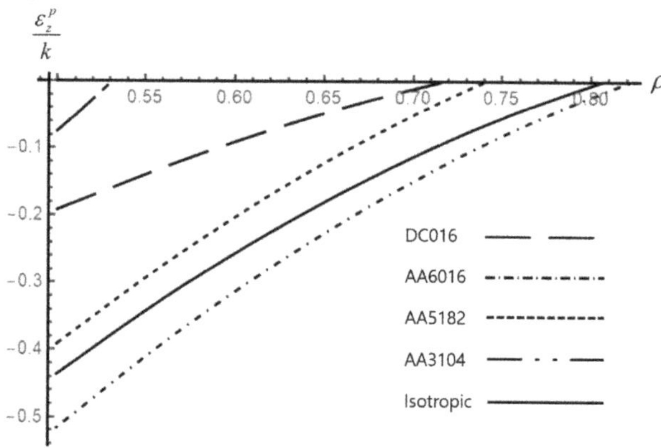

Figure 11. Effect of anisotropic properties on the distribution of the axial plastic strain in an $a = 0.5$ disk.

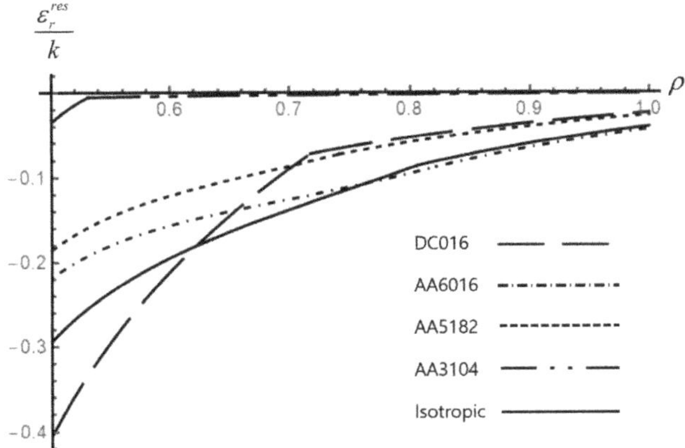

Figure 12. Effect of anisotropic properties on the distribution of the residual radial strain in an $a = 0.5$ disk.

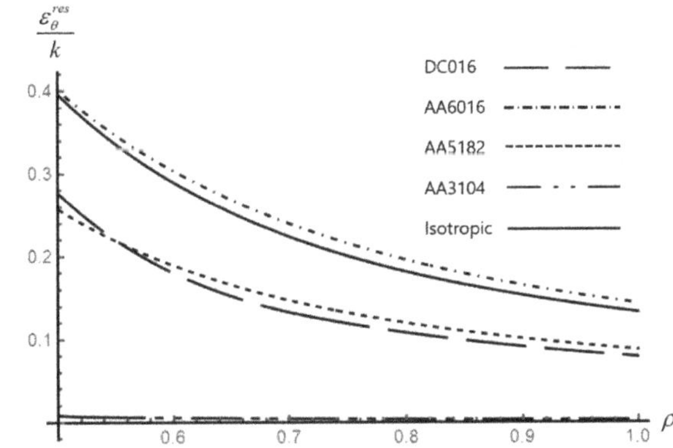

Figure 13. Effect of anisotropic properties on the distribution of the residual circumferential strain in an $a = 0.5$ disk.

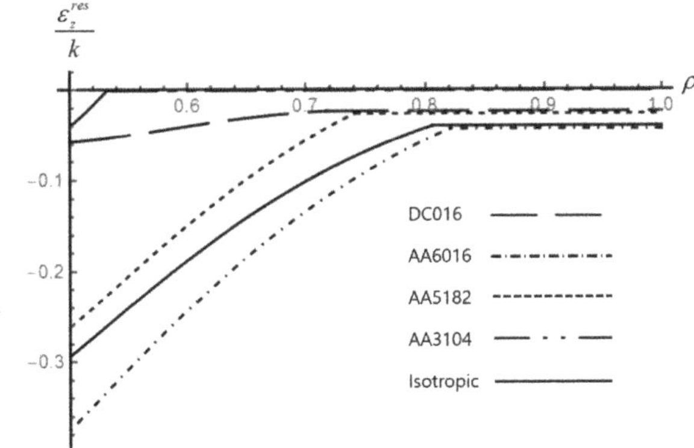

Figure 14. Effect of anisotropic properties on the distribution of the residual axial strain in an $a = 0.5$ disk.

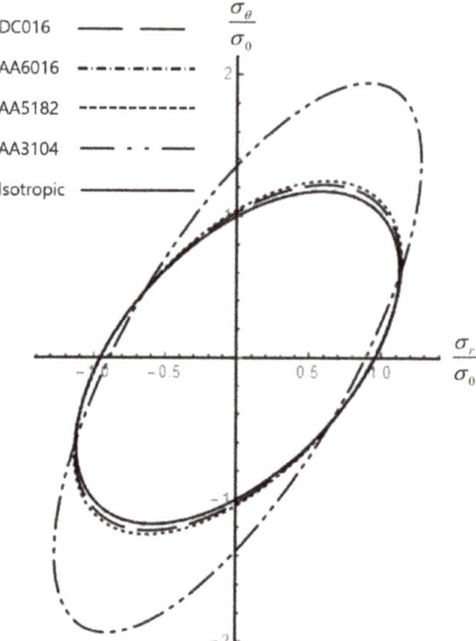

Figure 15. Yield loci for several materials.

6. Conclusions

A semi-analytic solution for the stresses and strains within a rotating elastic/plastic polar orthotropic annular disk was found under plane stress. The range of validity of plane stress solutions was determined in [33] by comparing such solutions with 3-D finite element solutions. The distribution of residual stresses and strains was determined as well. The constitutive equations consist of Hooke's law, the orthotropic yield criterion proposed in [14], and the associated flow rule. Therefore, the equations to be solved involve the strain rate tensor. This greatly adds to the difficulties of the solution as compared with the constitutive equations that relate the stresses to the strains (or allow for the strains to be immediately found by integrating relations between strain rate components). In order to facilitate numerical solution, the method developed in [11] was adopted. As a result, it is only necessary to use numerical methods for solving ordinary differential equations and to evaluate ordinary integrals, even though the solution depends on two independent variables—Ω and ρ.

It is seen from Equations (12), (21), (22), (32) and (33) that the parameter k introduced in Equation (8) is immaterial in the sense that scaling of any strain solution for a disk of given radius, Poisson's ratio, and anisotropic parameters provides the solutions for similar disks of material with the same Poisson's ratio and anisotropic parameters but any value of k.

It is known that numerical codes should be verified before their use in applications [34–36]. The present solution is useful for this purpose since ordinary differential equations can be solved numerically with a very high accuracy with no difficulty. In particular, comparison with the finite difference solution proposed in [10] was made. It was shown that the accuracy of the finite difference solution for the total circumferential strain at the inner radius of the disk may be insufficient (Table 2).

The solution found is for a rate-independent model of plasticity. However, in many cases, solutions for rate-dependent plasticity are required [37]. Moreover, disks of varying thickness and disks made of functionally graded materials are widely used in industry [38]. The general approach used in the present paper can be extended to at least some of these cases. This will be the subject of a subsequent investigation.

Author Contributions: All three authors participated in the research and in the writing of this paper.

Funding: This research received no external funding.

Acknowledgments: This work was performed while W. Jeong was a visiting research professor at Beihang University, Beijing, China.

Conflicts of Interest: The authors declare no conflict of interest.

References

1. Tahani, M.; Nosier, A.; Zebarjad, S.M. Deformation and stress analysis of circumferentially fiber-reinforced composite disks. *Int. J. Solids Struct.* **2005**, *42*, 2741–2754. [CrossRef]
2. Zare, H.R.; Darijani, H. Strengthening and design of the linear hardening thick-walled cylinders using the new method of rotational autofrettage. *Int. J. Mech. Sci.* **2017**, *124–125*, 1–8. [CrossRef]
3. Timoshenko, S.P.; Goodier, J.N. *Theory of Elasticity*, 3rd ed.; McGraw-Hill: New York, NY, USA, 1970.
4. Arnold, S.M.; Saleeb, A.F.; Al-Zoubi, N.R. Deformation and life analysis of composite flywheel disk systems. *Compos. Part B Eng.* **2002**, *33*, 433–459. [CrossRef]
5. Yildirim, V. Numerical/analytical solution to the elastic response of arbitrarily functionally graded polar orthotropic rotating discs. *J. Braz. Soc. Mech. Sci. Eng.* **2018**, *40*, 320. [CrossRef]
6. Eraslan, A.N. Inelastic deformations of rotating variable thickness solid disks by Tresca and von Mises criteria. *Int. J. Comput. Eng. Sci.* **2002**, *3*, 89–101. [CrossRef]
7. Hojjati, M.H.; Hassani, A. Theoretical and numerical analyses of rotating disks of non-uniform thickness and density. *Int. J. Press. Vessel. Pip.* **2008**, *85*, 694–700. [CrossRef]
8. Alexandrova, N.; Alexandrov, S.; Vila Real, P. Displacement field and strain distribution in a rotating annular disk. *Mech. Based Des. Struct.* **2004**, *32*, 441–457. [CrossRef]
9. Argeso, H. Analytical solutions to variable thickness and variable material property rotating disks for a new three-parameter variation function. *Mech. Based Des. Struct.* **2012**, *40*, 133–152. [CrossRef]
10. Alexandrova, N.N.; Real, P.M.M.V. Effect of anisotropy on stress-strain field in thin rotating disks. *Thin Walled Struct.* **2006**, *44*, 897–903. [CrossRef]
11. Lomakin, E.; Alexandrov, S.; Jeng, Y.-R. Stress and strain fields in rotating elastic/plastic annular discs. *Arch. Appl. Mech.* **2016**, *86*, 235–244. [CrossRef]
12. Alexandrov, S.; Chung, K.; Jeong, W. Stress and strain fields in rotating elastic/plastic annular disks of pressure-dependent material. *Mech. Based Des. Struct.* **2018**, *46*, 318–332. [CrossRef]
13. Prime, M.B. Amplified effect of mild plastic anisotropy on residual stress and strain anisotropy. *Int. J. Solids Struct.* **2017**, *118*, 70–77. [CrossRef]
14. Hill, R. *The Mathematical Theory of Plasticity*; Clarendon Press: Oxford, UK, 1950.
15. Singh, S.B.; Ray, S. Modeling the anisotropy and creep in orthotropic aluminum-silicon carbide composite rotating disc. *Mech. Mater.* **2002**, *34*, 363–372. [CrossRef]
16. Alexandrova, N.N.; Real, P.M.M.V. Singularities in a solution to a rotating orthotropic disk with temperature gradient. *Meccanica* **2006**, *41*, 197–205. [CrossRef]
17. Ari-Gur, J.; Stavsky, Y. On rotating polar-orthotropic circular disks. *Int. J. Solids Struct.* **1981**, *17*, 57–67. [CrossRef]
18. Liang, D.S.; Wang, H.J.; Chen, L.W. Vibration and stability of rotating polar orthotropic annular disks subjected to a stationary concentrated transverse load. *J. Sound Vib.* **2002**, *250*, 795–811. [CrossRef]
19. Koo, K.N. Vibration analysis and critical speeds of polar orthotropic annular disks in rotation. *Compos. Struct.* **2006**, *76*, 67–72. [CrossRef]
20. Peng, X.-L.; Li, X.-F. Elastic analysis of rotating functionally graded polar orthotropic disks. *Int. J. Mech. Sci.* **2012**, *60*, 84–91. [CrossRef]
21. Bert, C.W.; Paul, T.K. Failure analysis of rotating disks. *Int. J. Solids Struct.* **1995**, *32*, 1307–1318. [CrossRef]
22. You, L.H.; Zhang, J.J. Elastic-plastic stresses in a rotating solid disk. *Int. J. Mech. Sci.* **1999**, *41*, 269–282. [CrossRef]
23. Yahnioglu, N.; Akbarov, S.D. Stability loss analyses of the elastic and viscoplastic composite rotating thick circular plate in the framework of the three-dimensional linearized theory of stability. *Int. J. Mech. Sci.* **2002**, *44*, 1225–1244. [CrossRef]
24. Portnov, G.G.; Ochan, M.Y.; Bakis, C.E. Critical state of imbalanced rotating anisotropic disks with small radial and shear moduli. *Int. J. Solids Struct.* **2003**, *40*, 5219–5227. [CrossRef]

25. Çallıoğlu, H.; Topcu, M.; Tarakcılar, A.R. Elastic-plastic stress analysis of an orthotropic rotating disc. *Int. J. Mech. Sci.* **2006**, *48*, 985–990. [CrossRef]

26. Rees, D.W.A. A theory for swaging of discs and lugs. *Meccanica* **2011**, *46*, 1213–1237. [CrossRef]

27. Daghigh, V.; Daghigh, H.; Loghman, A.; Simoneau, A. Time-dependent creep analysis of rotating ferritic steel disk using Taylor series and Prandtl-Reuss relation. *Int. J. Mech. Sci.* **2013**, *77*, 40–46. [CrossRef]

28. Aleksandrova, N. Exact deformation analysis of a solid rotating elastic-perfectly plastic disk. *Int. J. Mech. Sci.* **2014**, *88*, 55–60. [CrossRef]

29. Alexandrov, S. *Elastic/Plastic Disks under Plane Stress Conditions*; Springer: New York, NY, USA, 2015.

30. Alexandrova, N.N.; Alexandrov, S. Elastic-plastic stress distribution in a plastically anisotropic rotating disk. *J. Appl. Mech.* **2004**, *71*, 427–429. [CrossRef]

31. Bouvier, S.; Teodosiu, C.; Haddadi, H.; Tabacaru, V. Anisotropic work-hardening behavior of structural steels and aluminium alloys at large strains. *J. Phys. IV France* **2003**, *105*, 215. [CrossRef]

32. Wu, P.D.; Jain, M.; Savoie, J.; MacEwen, S.R.; Tugcu, P.; Neale, K.W. Evaluation of anisotropic yield functions for aluminum sheets. *Int. J. Plast.* **2003**, *19*, 121–138. [CrossRef]

33. Kammal, S.M.; Dixit, U.S.; Roy, A.; Liu, Q.; Silberschmidt, V.V. Comparison of plane-stress, generalized-plane-strain and 3D FEM elastic-plastic analyses of thick-walled cylinders subjected to radial thermal gradient. *Int. J. Mech. Sci.* **2017**, *131–132*, 744–752. [CrossRef]

34. Roberts, S.M.; Hall, F.R.; Bael, A.V.; Hartley, P.; Pillinger, I.; Sturgess, C.E.N.; Houtte, P.V.; Aernoudt, E. Benchmark tests for 3-D, elasto-plastic, finite-element codes for the modelling of metal forming processes. *J. Mater. Process. Technol.* **1992**, *34*, 61–68. [CrossRef]

35. Becker, A.A.; Hyde, T.H.; Sun, W.; Andersson, P. Benchmarks for finite element analysis of creep continuum damage mechanics. *Comp. Mater. Sci.* **2002**, *25*, 34–41. [CrossRef]

36. Helsing, J.; Jonsson, A. On the accuracy of benchmark tables and graphical results in the applied mechanics literature. *J. Appl. Mech.* **2002**, *69*, 88–90. [CrossRef]

37. Zharfi, H.; EkhteraeiToussi, H. Time dependent creep analysis in thick FGM rotating disk with two-dimensional patterns of heterogeneity. *Int. J. Mech. Sci.* **2018**, *140*, 351–360. [CrossRef]

38. Yildirim, V. A parametric study on the centrifugal force-induced stress and displacements in power-law graded hyperbolic discs. *Lat. Am. J. Solids Strut.* **2018**, *15*, e34.

Article

Investigation on the Effect of Type of Cooling on the Properties of Aluminum Alloy during Warm/Hot Hydromechanical Deep Drawing

Gaoshen Cai [1,*], Chuanyu Wu [1] and Dongxing Zhang [2]

[1] Faculty of Mechanical Engineering & Automation, Zhejiang Sci-Tech University, Hangzhou 310018, China; cywu@zstu.edu.cn
[2] Department of Mechanical & Material Engineering, Western University, London, ON N6A 5B9, Canada; zhangdx1113@gmail.com
* Correspondence: caigaocan@zstu.edu.cn

Received: 9 July 2018; Accepted: 24 August 2018; Published: 26 August 2018

Abstract: The warm sheet cylindrical deep drawing experiment of aluminum alloy was carried out and macro-mechanical properties and microstructure evolution of hydro-formed cups with different cooling medium were analyzed, which aimed to investigate the effects of different types of cooling on mechanical properties and microstructure of cylindrical cups hydro-formed by warm Hydro-mechanical Deep Drawing (HDD). Results show that, under the condition of warm hydroforming, the mechanical properties such as yield stress and ultimate strength were influenced very little by air or water cooling. Grain coarsening of these hydro-formed cups can be inhibited to a certain extent with subsequent rapid water cooling. Moreover, it shows that the processing with warm sheet hydroforming and subsequent rapid cooling of 7075-O aluminum alloy has a positive significance in maintaining the stability of macro mechanical properties and inhibiting the degradation of the microstructure of materials.

Keywords: hydro-mechanical deep drawing (HDD); mechanical property; type of cooling; microstructure

---

1. Introduction

Warm/hot sheet hydroforming has prominent advantages in improving the forming limit due to combining the advantages of both cold sheet hydroforming and warm sheet forming [1,2]. Since temperatures and fluid pressures can lead to more formability than by using cold sheet hydroforming or warm sheet forming, a warm/hot sheet hydroforming process is suitable to form the thin-wall structural parts with complex surfaces of hard-to-form materials at Room Temperature (RT), which can solve the manufacturing problems in hard shaping or hard integral shaping using conventional forming methods [3,4]. Like cold sheet hydroforming, according to the action of fluid pressure in the forming process, warm/hot sheet hydroforming is classified into two styles: passive forming and active forming, which is shown in Figure 1 [5]. When forming parts like cylindrical cups, the fluid pressure plays a supporting role in the forming process, which can be called HDD, and the passive forming style can be used. By contrast, the other style can be chosen when more active fluid pressures are needed for forming parts, which can be referred to as warm/hot sheet hydro-bulging.

Even though the plasticity of materials can be improved in a warm/hot environment, it will inevitably lead to the deterioration tendency of material properties due to microstructure variation. In addition, when encountering a reduced temperature gradient at an elevated temperature, the microstructure deterioration tendency of metal materials may be inhibited and transformed in a favorable direction [6,7]. From this, the corresponding solution to the microstructure property deterioration of materials can be determined.

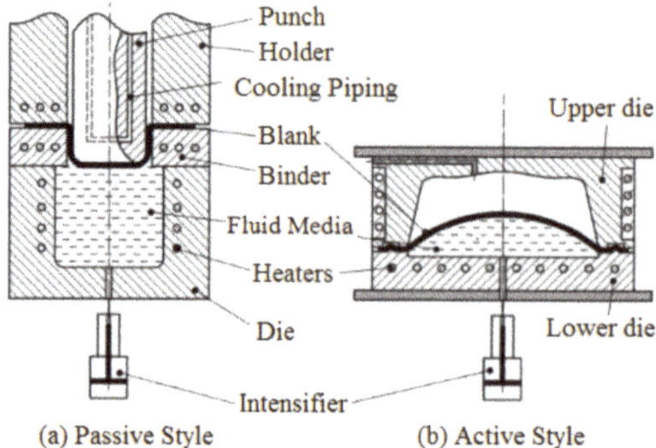

Figure 1. Schematic of warm sheet hydroforming.

Unlike steel materials, the strengthening process of heat treatable aluminum alloys (such as 7075) is precipitation dispersion strengthening. It requires the rapid cooling of the materials from the solution temperature to room temperature rapidly and undergoes natural aging or artificial aging to strengthen its matrix due to precipitating the dispersed phase. Some relevant studies indicated [8–10] that the hydroforming technology process under a higher heat environment requires materials to have a certain strength and the forming temperature should not be too high in contrast to those with a warm or hot forming type. The hard-to-form materials, which cannot be formed using a traditional forming method, can be formed using warm/hot sheet hydroforming with subsequent rapid cooling to proceed incomplete solution strengthening. Then it can maintain the stability of macro-mechanical properties and inhibit the degradation of the microstructure of materials to a certain extent. In addition, research shows that [5,11,12] assuming that the material is isotropy, stress-strain curves obtained by the bulging test have been very close to the actual flow properties of materials and can reflect the stress-strain state during the forming process. Considering that the cylindrical cups HDD experiment is a classic basic research method [13,14], it was used to research the effect of the process of warm/hot sheet hydroforming with subsequent rapid cooling on mechanical properties and microstructure evolution of the cups, which were formed by warm/hot sheet HDD [15,16]. Meanwhile, the isotropic material was assumed in this study.

As mentioned above, using a special warm/hot sheet hydrobulging-hydromechanical deep drawing equipment, a warm/hot sheet HDD experiment of 7075-O aluminum alloy was carried out and macro mechanical properties and microstructure evolution of hydro-formed cups with different cooling medium (which represents different types of cooling) were researched, which aimed to reveal the influence rule of different types of cooling on mechanical properties and microstructure of cylindrical cups formed by warm/hot sheet hydroforming in this study. In addition, it is of great significance to further study the process characteristics of warm sheet hydroforming and inhibit the degradation of the microstructure of materials [17]. Moreover, the results can promote the application process of warm/hot sheet hydroforming technology effectively in the field of aerospace.

2. Warm/Hot Sheet HDD Experiment

2.1. Experimental Material

The HDD testing material used was an AA 7075-O aluminum alloy sheet with a thickness of 1 mm, which was produced by Alcoa (Al-Zn-Mg-Cu series high-strength alloy). As a representative alloy used widely in the aerospace field and as a hard-to-form material at room temperature, it was selected to conduct the hydro-bulging test. Table 1 shows the chemical composition [18] and Figure 2 shows the initial microstructure of 7075-O aluminum alloy obtained using the Scanning Electron Microscope (SEM).

Table 1. Chemical composition of 7075-O aluminum alloy (wt %) [18].

Composition	Zn	Mg	Cu	Mn	Cr	Fe	Si	Ti	others	Al
Percent (%)	5.1	2.1	1.2	0.3	0.18	0.5	0.4	0.2	0.2	rest

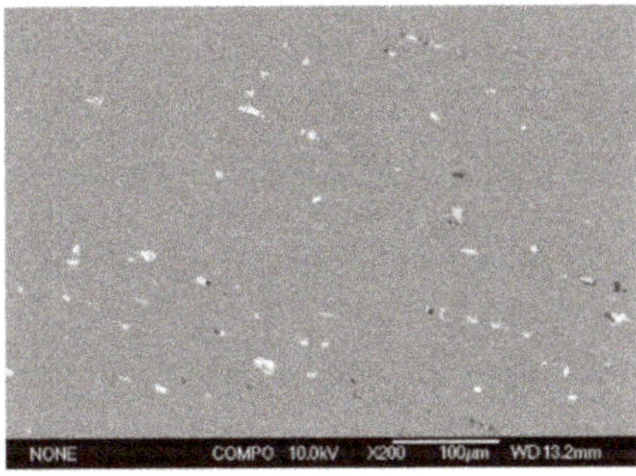

Figure 2. The initial microstructure of AA 7075-O aluminum alloy captured with the Scanning Electron Microscope (SEM).

2.2. Test Equipment

The HDD test equipment used was the self-development YRJ-50 machine [6,18,19], by which not only can the stress-strain curve be confirmed and obtained by a warm bulging test [19,20] but also a warm sheet HDD test can be carried out. It consists of a mechanical body, a detecting system, and a control system. Figure 3 shows the equipment and the setup of this HDD test.

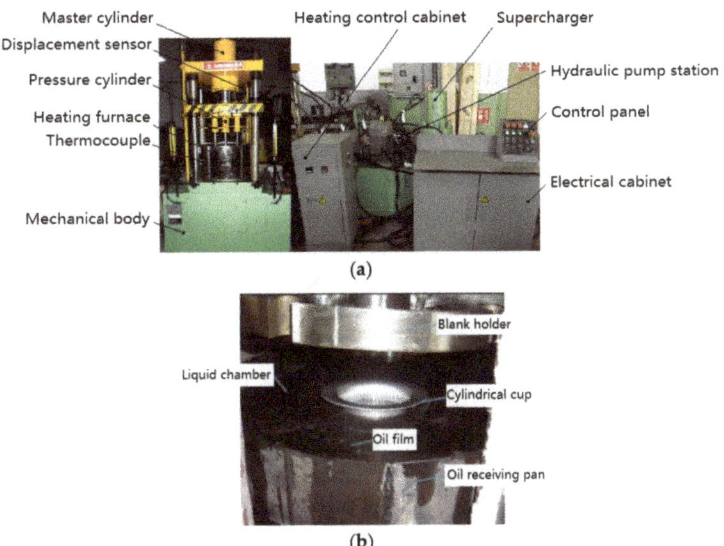

Figure 3. Equipment and setup of HDD test. (**a**) Warm sheet hydro-bulging-hydro-mechanical deep drawing equipment; and (**b**) experimental setup of warm hydro-mechanical deep drawing.

2.3. Experimental Results

At RT and 210 °C, the warm sheet HDD experiment of the aluminum alloy 7075-O was carried out using the equipment mentioned above. In addition, the test conditions include: the punch diameter was 80 mm, the punch nose radius was 10 mm, the draw die shoulder radius was 10 mm, the aperture of the blank holder was 81.5 mm, the blank holder radius was 8 mm, the inlet diameter of the liquid chamber was 85 mm, the blank diameter was 200 mm (210 °C and 160 mm (RT), the final diameter of the cup after forming in theory was nearly 81 mm, and the deep drawing depth was 85 mm. As a blank holder force, liquid pressure-punch stroke curves and liquid pressure-punch stroke curves are very important influencing factors during the process of warm sheet HDD. They were designed before testing, which is shown in Figure 4a,b. After being formed, air cooling and water cooling were selected to cool the test samples to RT rapidly (the processing rate is shown in Figure 4c. Then the testing samples were obtained. Figure 5 shows the cylindrical cups hydro-formed by a warm sheet HDD at 210 °C.

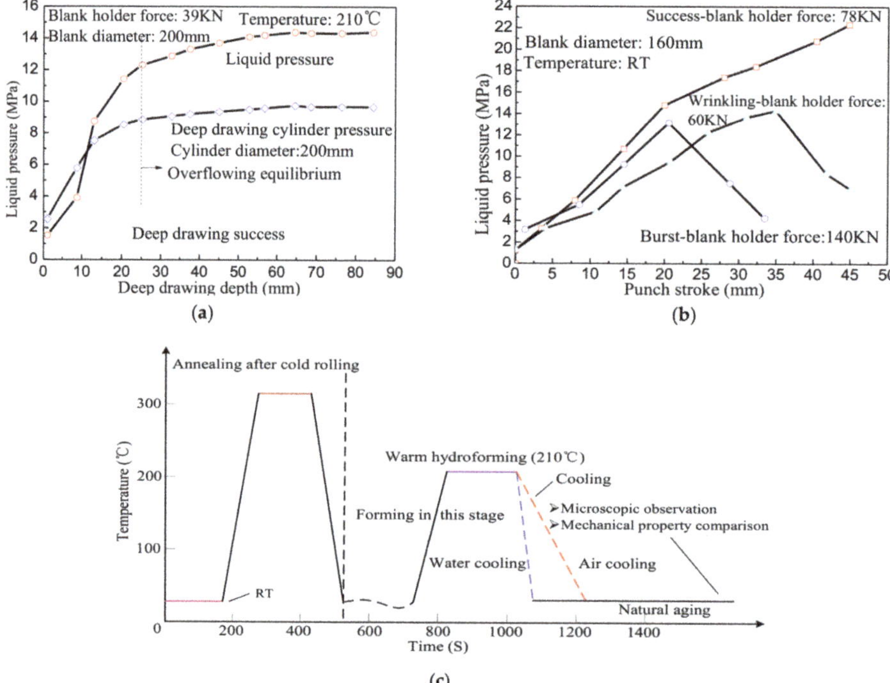

Figure 4. Experimental scheme of warm sheet HDD experiment. (**a**) liquid pressure-punch stroke curve and drawing force-punch stroke curve at 210 °C. (**b**) curves of liquid pressure-punch stroke in HDD at RT; and (**c**) processing route of the warm sheet HDD experiment.

Figure 5. Cylindrical cups hydro-formed by a warm sheet HDD at 210 °C.

3. Performance Evaluation of Aluminum Alloy Cylindrical Cups Hydro-Formed by Warm Sheet HDD

3.1. Evaluation Testing Scheme

After being formed by warm/hot HDD, as the deformation of sheet materials has been accumulated, the microstructure of test sheets is in an unstable state to a certain extent. Different types of cooling (different cooling medium) represent different cooling rates in order to evaluate the macro performance of aluminum alloy cylindrical cups. The uniaxial tension samples in the straight wall of cups with the rolling direction of 0 degrees were sampled. Then, in order to evaluate their mechanical performance parameters that yield strength, tensile strength, and elongation, the uniaxial tension experiment at RT was conducted. On the other hand, with the rolling direction of 90 degrees and clipping the narrow strips with the width of 10 mm from the flange to the bottom of cylindrical cups, the microstructure observation samples were prepared (see Figure 6). There is a different deformation due to a different deformation area since both flange and a cup wall are the parts with large deformation during the warm/hot sheet HDD process of aluminum alloy cylindrical cups. The two positions of cups were selected to observe their microstructure variation. The performance measurement and evaluation plan of formed cylindrical cups is shown in Table 2.

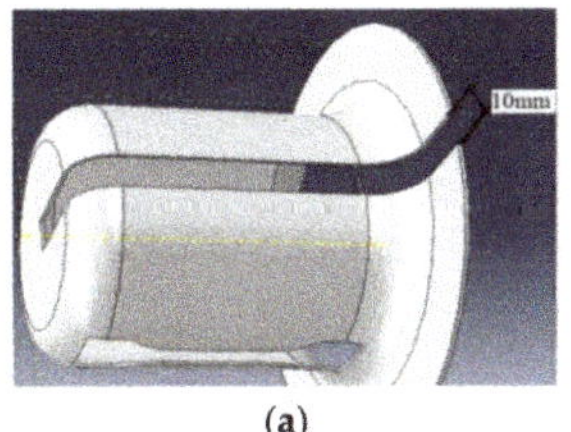
(a)

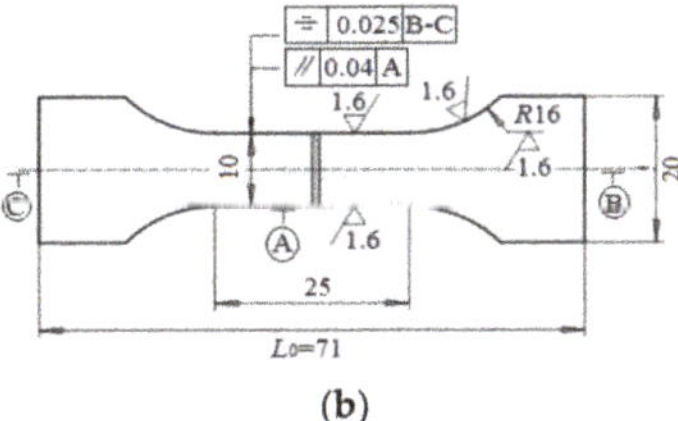
(b)

Figure 6. Schematic diagram of specimens sampling for uniaxial tension and metallographic observation. (**a**) sampling area for uniaxial tension and metallographic observation; and (**b**) sample size for uniaxial tension.

Table 2. Experimental plan for mechanical properties after warm sheet HDD.

Serial Number	Temperature (°C)	Maximum Pressure (MPa)	Sizes of Cylindrical Cups (mm)	Type of Cooling	Samples of Uniaxial Tension (mm)	Rolling Direction (°)
1	210	$P_{max} = 14.5$	$t_0 = 1.0; \varphi = 80; h = 85$	Water cooling	$L_0 = 71$	0
2	210	$P_{max} = 14.4$	$t_0 = 1.0; \varphi = 80; h = 85$	Water cooling	$L_0 = 71$	0
3	210	$P_{max} = 14.3$	$t_0 = 1.0; \varphi = 80; h = 85$	Water cooling	$L_0 = 71$	0
4	210	$P_{max} = 14.5$	$t_0 = 1.0; \varphi = 80; h = 85$	Air cooling	$L_0 = 71$	0
5	210	$P_{max} = 14.4$	$t_0 = 1.0; \varphi = 80; h = 85$	Air cooling	$L_0 = 71$	0
6	210	$P_{max} = 14.3$	$t_0 = 1.0; \varphi = 80; h = 85$	Air cooling	$L_0 = 71$	0

3.2. Effect of Type of Cooling on Mechanical Properties of Testing Samples

In order to observe the influence of different types of cooling on mechanical properties of cylindrical cups after being formed by warm/hot sheet HDD, it was necessary to carry out the uniaxial tension test, which was conducted using the electro-hydraulic servo tension test machine Instron 8801 (in Figure 7) on the evaluation samples. During testing, two samples in each group were selected to obtain the average values, which yielded the test results. Then the following equation was used.

$$L_0 = 5.65\sqrt{S_0} \tag{1}$$

where S_0 is the minimum value of the section of samples and L_0 is the initial reference length for calculating the percentage elongation, which is the sample length for uniaxial tension. This is shown in Figure 6b. Equation (1) describes the relation between L_0 and S_0, which means the test sample is a scaling sample conforming to the international common standard. Then the reliable and significant testing results can be insured using this method.

Figure 7. Experimental setup of a uniaxial tension test.

According to the performance measurement scheme, the mechanical property test results of cylindrical cups are shown in Figure 8a. It can be indicated that the yield strength is 224.18 MPa and the ultimate strength is 260.43 MPa under the condition of air cooling while, under the condition of water cooling, the yield strength is 228.43 MPa and the ultimate strength is 260.66 MPa. Results show that the yield strength increases a little under water cooling compared with air cooling and the ultimate strength is nearly unchanged during either type of cooling. It indicates that different types of cooling have little effect on the mechanical properties of 7075-O aluminum alloy cylindrical cups formed using warm/hot sheet HDD. Figure 8b shows the results of specific elongation after cooling using two different types of cooling. The value with air cooling is 11.83% and, with water cooling, is 11.18%. Results show that the specific elongation of cylindrical cups under water cooling decreases a little more than that of air cooling. In general, though, the effect of different types of cooling on the specific elongation is not clear.

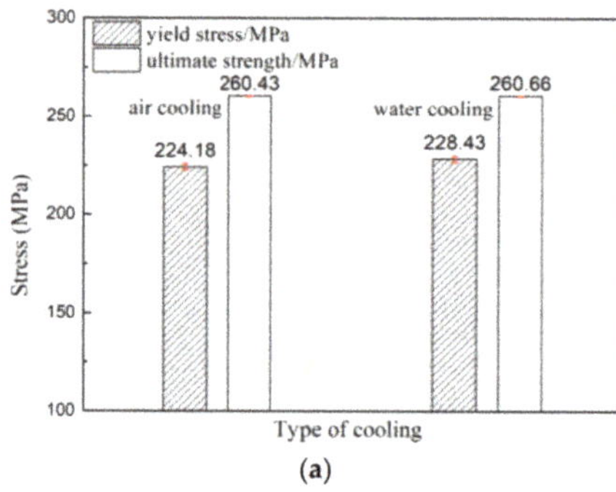
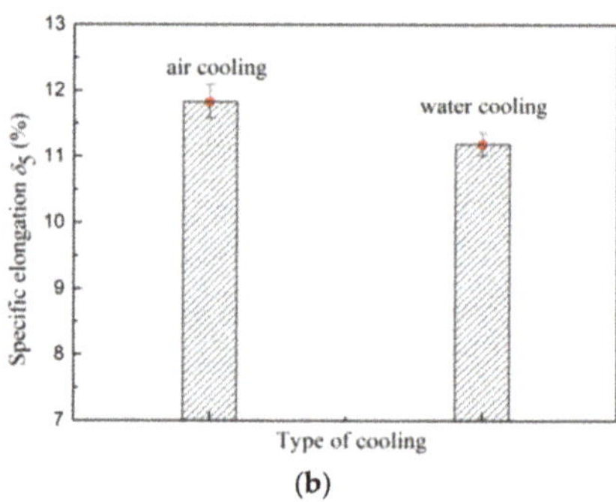

Figure 8. The influence of different types of cooling on yield stress, ultimate strength, and specific elongation. (**a**) Yield stress and ultimate strength; (**b**) Specific elongation.

In theory, the different types of cooling mainly affects the cooling rate of samples during the forming process of materials using warm/hot sheet hydroforming. Due to the result of cylindrical cups after being formed, the influence on mechanical properties may not be clear. It shows that the influence of different types of cooling on properties of aluminum alloys may be in a macro field rather than in a micro field, which is demonstrated by the experimental results described above. These results may be positively significant for maintaining the stability of macro mechanical properties of sheet metals.

3.3. Effect of Type of Cooling on the Microstructure of Testing Samples

The metallographic test and the Electron Back-Scattered Diffraction (EBSD) analysis on flanges and cup walls of samples after being formed using different types of cooling were carried out to observe the microstructure variation of cylindrical cups hydro-formed by warm/hot sheet HDD. According to GB/T 3246.1-2000, test samples and the special parts of the aluminum alloy cylindrical cups were corroded using a Keller reagent (1 mL HF, 1.5 mL HCl, 2.5 mL HNO_3, and 95 mL H_2O). Then, using the optical microscope Axiovert 200MAT produced by the Zeiss company in Oberkochen of Germany, the samples were observed. The results are shown in Figure 9.

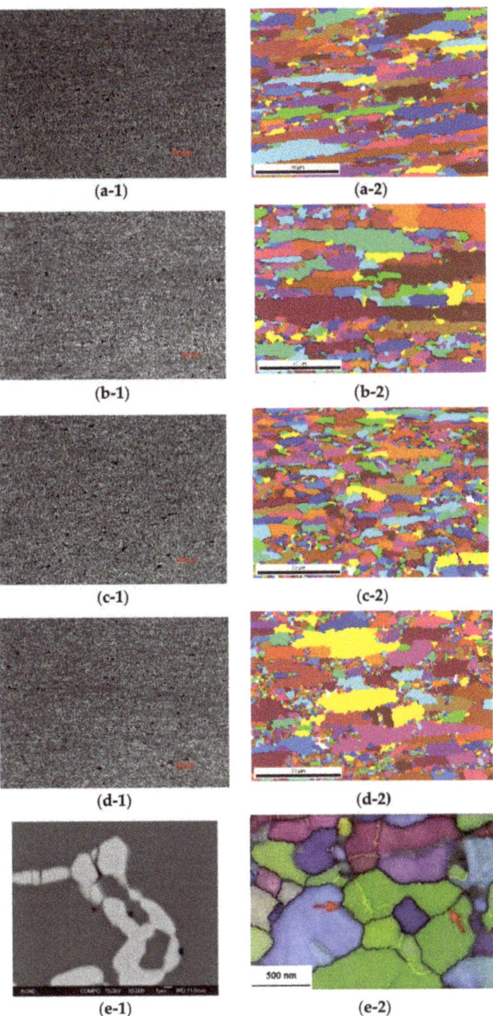

Figure 9. The influence of different types of cooling on microstructures. (**a-1**) The microstructure of flange by a metallographic test with air cooling. (**a-2**) The microstructure of flange by EBSD with air cooling. (**b-1**) The microstructure of cup wall by a metallographic test with air cooling. (**b-2**) The microstructure of cup wall by EBSD with air cooling. (**c-1**) The microstructure of flange by a metallographic test with water cooling. (**c-2**) The microstructure of flange by EBSD with water cooling. (**d-1**) The microstructure of the cup wall by a metallographic test with water cooling. (**d-2**) The microstructure of a cup wall by EBSD with water cooling. (**e-1**) The microstructure of flange by SEM with air cooling. (**e-2**) Orientation image of a cup wall by EBSD with water cooling.

Studies have shown that [21–23], under the condition of a warm/elevated temperature, grain growth is mainly caused by plastic deformation and it is a dynamic recovery process. As is shown in Figure 9(a-1,c-1,e-1), after being formed using warm/hot sheet HDD, the compounds in flange of 7075-O aluminum alloy was permutated along the calendaring direction after fracture. In addition, the phase particle was precipitated in the α (Al) matrix, which was shown clearly in Figure 9(e-1). Figure 9(b-1,d-1) show that the compounds in a cylindrical cup wall further fractured, which were a stronger permutation along the deformation direction than that of flange and there was a dispersed phase in the α (Al) matrix. According to References [21–23], on the whole, there were more oversized second phase particles in the metallographic structure under the two types of cooling and these second phases were different in morphology including the acicular metastable phase and the punctiform stable phase [21,22]. From Figure 9(a-2) to Figure 9(e-2), they show the results of grain morphology observed using EBSD technology. Figure 9(e-2) shows the orientation image of a cup wall with water cooling from which it can be seen that, along the **calendaring** direction, the grain was strip, which illustrates the grain anisotropy of formed cylindrical cups in a **calendaring** direction and vertical direction [23].

Comparing Figure 9(a-2) with (b-2) and Figure 9(c-2) with (d-2), it can be seen that, under the same types of cooling, the grain size in the cylindrical cup wall is larger than that of the flange. The reason is that, during the process of deep drawing, the flange was acted upon by two forces of a pressure-tensile stress (pressure stress of circumferential and thickness direction). Besides the thickness normal stress provided by hydraulic pressure, the cylindrical cup wall was also acted upon by the two forces of tensile stress.

The variation of grain size under different types of cooling are shown in Figure 10. Under the condition of air cooling, the maximum grain size in the flange was 34 μm, which accounts for 13.3% of the sampling area. The maximum grain size in the cylindrical cup wall was 45 μm, which accounts for 19.7%. While under the condition of water cooling, the maximum grain size in the flange was 20 μm and it accounts for 8.3% of the sampling area. The maximum grain size of the cylindrical cup wall was 37 μm and it accounts for 9.4%. It shows that, with the air cooling condition, there were an increase in the oversized grain. By contrast, there was grain coarsening to a certain extent with a water cooling condition, but the grain size was uniform on the whole and the grain size in the flange and cylindrical cup wall was smaller than that of air cooling. In addition, the maximum grain size appeared in the same cylindrical cup wall under both of the types of cooling while the maximum grain size under the water cooling condition was about 80% than that of air cooling. The results show that the processing warm sheet hydroforming with thereafter subsequent cooling of 7075-O aluminum alloy can inhibit the grain coarsening to a certain extent, which proves that it is of clear positive significance in maintaining the stability of macro mechanical properties and inhibiting the degradation of the microstructure of materials.

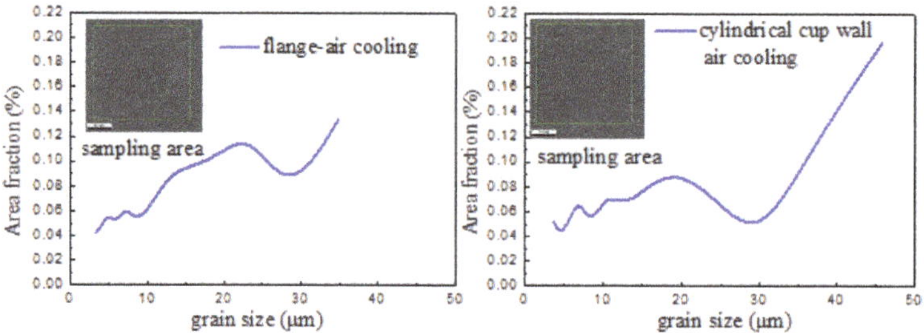

Figure 10. *Cont.*

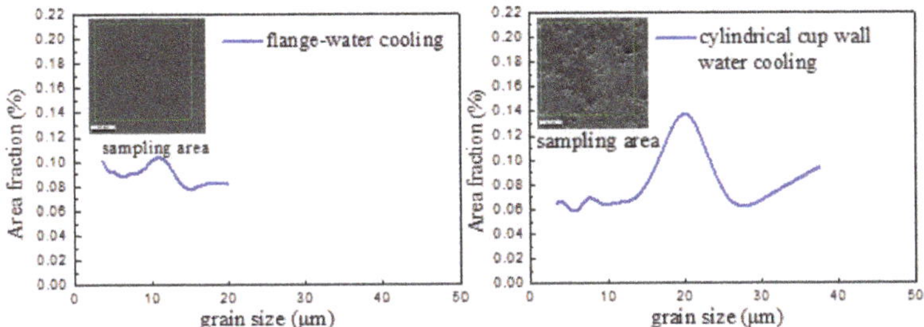

Figure 10. The influence of different types of cooling on the grain size.

4. Conclusions

This paper brings new experimental results concerning the effects of a type of cooling on the properties of aluminum alloy using a warm/hot sheet hydroforming process. Warm/hot sheet HDD experimentation was carried out to investigate the effects of different types of cooling on mechanical properties and microstructure evolution of cylindrical cups. The main results obtained in this study can be summarized by the following points below.

1. It shows that, under the condition of warm hydroforming, the mechanical properties of the 7075-O aluminum alloy cylindrical cups were influenced very little by different types of cooling. Compared with air cooling, there were more precipitates of the cups with water cooling, but the ultimate strength was nearly unchanged. While the yield strength increased slightly and the specific elongation tended to decrease a little under the condition of water cooling.
2. Under the condition of air cooling, the grain of the flange and the cylindrical cup wall of the formed cups were coarsened inordinately and the grain of the cylindrical cup wall was the most serious in which the maximum grain size was 45 μm. Alternately, under the condition of water cooling, the grain size of the flange and the cylindrical cup wall of the cups were inhibited effectively and the grain size was smaller and more uniform than that of air cooling.
3. It proves that the grain coarsening of the 7075-O aluminum alloy hydro-formed cups can be inhibited to a certain extent during warm/hot sheet hydroforming with subsequent rapid water cooling, which indicates that there is a positive significance in maintaining the stability of macro mechanical properties and inhibiting the degradation of the materials' microstructure.

Author Contributions: D.Z. analyzed the experimental data. G.C. analyzed the date and wrote this paper. C.W. proposed the ideas and reviewed the paper.

Funding: This research received no external funding.

Acknowledgments: This work was supported by the Zhejiang Provincial Natural Science Foundation (ZJNSF) with Grant No. LQ18E050010 and the Zhejiang Sci-Tech University Research Foundation with Grant No. 17022073-Y.

References

1. Lang, L.H.; Liu, B.S.; Liu, H.J.; Lyamina, E. Simulation of aluminium alloy 5A06 warm/hot hydromechanical sheet deep drawing. *Mater. Sci. Forum.* **2009**, *623*, 61–69. [CrossRef]
2. Lang, L.H.; Wang, Z.R.; Kang, D.C.; Yuan, S.J.; Zhang, S.H.; Danckert, J.; Nielsen, K.B. Hydroforming highlights: Sheet hydroforming and tube hydroforming. *J. Mater. Process. Technol.* **2004**, *151*, 165–177. [CrossRef]
3. Liu, B.S.; Lang, L.H.; Zeng, Y.S.; Lin, J.G. Forming characteristic of sheet hydroforming under the influence of through-thickness normal stress. *J. Mater. Process. Technol.* **2012**, *212*, 1875–1884. [CrossRef]

4. Lang, L.H.; Cai, G.S.; Liu, K.N.; Alexandrov, S.; Du, P.M.; Zheng, H. Investigation on the effect of through thickness normal stress on forming limit at elevated temperature by using modified M-K model. *Int. J. Mater. Form.* **2015**, *8*, 211–228. [CrossRef]

5. Lang, L.H.; Liu, B.S.; Li, T.; Zhao, X.N.; Zeng, Y.S. Experimental investigation on hydromechanical deep drawing of aluminum alloy with heated media. *Steel Res. Int.* **2012**, *83*, 230–237. [CrossRef]

6. Cai, G.S.; Wu, C.Y.; Gao, Z.P.; Lang, L.H.; Alexandrov, S. Research on Al-alloy sheet forming formability during warm/hot sheet hydroforming based on elliptical warm bulging test. *AIP Adv.* **2018**, *8*, 1–9. [CrossRef]

7. Lang, L.H.; Du, P.M.; Liu, B.S.; Cai, G.S.; Liu, K.N. Pressure rate controlled unified constitutive equations based on microstructure evolution for warm hydroforming. *J. Alloys Compd.* **2013**, *574*, 41–48. [CrossRef]

8. Liu, K.N.; Lang, L.H.; Cai, G.S.; Yang, X.Y.; Guo, C.; Liu, B.S. A novel approach to determine plastic hardening curves of AA7075 sheet utilizing hydraulic bulging test at elevated temperature. *Int. J. Mech. Sci.* **2015**, *100*, 328–338. [CrossRef]

9. Kaya, S.; Altan, T.; Groche, P.; Klöpsch, C. Determination of the flow stress of magnesium Az31-O sheet at elevated temperatures using the hydraulic bulge test. *Int. J. Mach. Tools Manuf.* **2008**, *48*, 550–557. [CrossRef]

10. Mahabunphachai, S.; Koc, M. Investigations on forming of aluminum 5052 and 6061 sheet alloys at warm temperatures. *Mater. Des.* **2010**, *31*, 2422–2434. [CrossRef]

11. Jeong, Y.; Gnäupel-Herold, T.; Barlat, F.; Iadicola, M.; Creuziger, A.; Lee, M.-G. Evaluation of biaxial flow stress based on elasto-viscoplastic self-consistent analysis of X-ray diffraction measurements. *Int. J. Plast.* **2015**, *66*, 103–118. [CrossRef]

12. Min, J.Y.; Stoughton, T.B.; Carsley, J.E.; Carlson, B.E.; Lin, J.P.; Gao, X.L. Accurate characterization of biaxial stress-strain response of sheet metal from bulge testing. *Int. J. Plast.* **2017**, *94*, 192–213. [CrossRef]

13. Groche, P.; Huber, R.; Doerr, J.; Schmoeckel, D. Hydromechanical deep-drawing of aluminium-alloys at Elevated Temperatures. *CIRP Ann.* **2002**, *51*, 215–218. [CrossRef]

14. Zafar, R.; Lang, L.H.; Zhang, R.J. Analysis of hydro-mechanical deep drawing and the effects of cavity pressure on quality of simultaneously formed three-layer Al alloy parts. *Int. J. Adv. Manuf. Tech.* **2015**, *80*, 2117–2128. [CrossRef]

15. He, D.G.; Lin, Y.C.; Chen, J.; Chen, D.D.; Huang, J.; Tang, Y.; Chen, M.S. Microstructural evolution and support vector regression model for an aged Ni-based superalloy during two-stage hot forming with stepped strain rates. *Mater. Des.* **2018**, *154*, 51–62. [CrossRef]

16. Ji, H.C.; Liu, J.P.; Wang, B.Y.; Tang, X.F.; Lin, J.G.; Huo, Y.M. Microstructure evolution and constitutive equations for the high-temperature deformation of 5Cr21Mn9Ni4N heat-resistant steel. *J. Alloys Compd.* **2017**, *693*, 674–687. [CrossRef]

17. Lin, Y.C.; Luo, S.C.; Yin, L.X.; Huang, J. Microstructural evolution and high temperature flow behaviors of a homogenized Sr-modified Al-Si-Mg alloy. *J. Alloys Compd.* **2018**, *739*, 590–599. [CrossRef]

18. Cai, G.S.; Zhou, X.J.; Lang, L.H.; Alexandrov, S. Research on aluminum alloy sheet thermoplastic deformation behavior based upon warm bulging test. *AIP Adv.* **2016**, *6*, 1–8. [CrossRef]

19. Cai, G.S.; Lang, L.H.; Liu, K.N.; Alexandrov, S.; Zhang, D.X.; Yang, X.Y.; Guo, C. Research on the effect of flow stress calculation on aluminum alloy sheet deformation behavior based on warm bulging test. *Met. Mater. Int.* **2015**, *21*, 365–373. [CrossRef]

20. Koç, M.; Billur, E.; Necati, Ö.N. An experimental study on the comparative assessment of hydraulic bulge test analysis methods. *Mater. Des.* **2011**, *32*, 272–281. [CrossRef]

21. Hu, J.Q. Experimental Simulation of Control Rolling Process of 7075 Aluminum Alloy. Master's Thesis, Central South University, Changsha, China, 2003.

22. Lu, J.; Song, Y.L.; Hua, L.; Zheng, K.L.; Dai, D.G. Thermal deformation behavior and processing maps of 7075 aluminum alloy sheet based on isothermal uniaxial tensile tests. *J. Alloys Compd.* **2018**, *767*, 856–869. [CrossRef]

23. Tian, W.M.; Li, S.M.; Liu, J.H.; Yu, M.; Du, Y.J. Preparation of bimodal grain size 7075 aviation aluminum alloys and their corrosion properties. *Chin. J. Aeronaut.* **2017**, *30*, 1777–1788. [CrossRef]

Article

Numerical Study of Dynamic Properties of Fractional Viscoplasticity Model

Michał Szymczyk [1], Marcin Nowak [2] and Wojciech Sumelka [1,*]

[1] Institute of Structural Engineering, Poznań University of Technology, Piotrowo 5 street, 60-965 Poznań, Poland; michal.g.szymczyk@doctorate.put.poznan.pl

[2] Institute of Fundamental Technological Research, Polish Academy of Sciences, Pawińskiego 5B street, 02-106 Warsaw, Poland; nowakm@ippt.pan.pl

* Correspondence: wojciech.sumelka@put.poznan.pl

Received: 17 June 2018; Accepted: 10 July 2018; Published: 13 July 2018

Abstract: The *fractional viscoplasticity* (FV) concept combines the Perzyna type viscoplastic model and fractional calculus. This formulation includes: (i) rate-dependence; (ii) plastic anisotropy; (iii) non-normality; (iv) directional viscosity; (v) implicit/time non-locality; and (vi) explicit/stress-fractional non-locality. This paper presents a comprehensive analysis of the above mentioned FV properties, together with a detailed discussion on a general 3D numerical implementation for the explicit time integration scheme.

Keywords: fractional viscoplasticity; rate dependence; plastic anisotropy; non-normality; directional viscosity; explicit/implicit non-locality.

1. Introduction

In search of a generalization of existing models describing experimentally observed phenomena, the concept of fractional calculus [1,2] emerged as a tool that in the recent years became widely applied. Among the areas in which this theory has found application, it is worth mentioning mechanics where one can distinguish: (i) time-fractional models; (ii) space-fractional models; and (iii) stress-fractional models. For example, in [3], the time-fractional model was used to describe the time-dependent mechanical property evolution in ductile metals. The fractional oscillators were analyzed in [4], whereas the heat and mass transfer analysis in the framework of fractional calculus was presented in [5,6]. Furthermore, the analysis and modeling of turbulent flow in a porous medium [7], fluid transport induced by the osmotic pressure of glucose and albumin [8], wave propagation in the viscoelastic material [9], non-local boundary value problem [10], and evolution for the damage variable for hyperelastic materials [11] with an application of time-fractional derivative suggests great versatility of this approach. On the other hand, the space-fractional models are successfully used in mechanics to describe the deformation of a harmonic oscillator [12], deformation of an infinite bar subjected to a self-equilibrated load distribution [13], modeling plane strain and plane stress elasticity [14], Euler–Bernoulli beam [15], Darcy's flow in porous media [16] and fractional strain formulation [17]. Finally, the stress-fractional models [18,19] and their finite element implementations were used to study the granular soils under drained cyclic loading [20], and monotonic triaxial compression [21]. Concluding, one should emphasize that regardless of the specific formulation, the fractional operators have one common feature, namely, the 'change' of a selected variable is based on integration over a closed interval, thus extending the definition of integer order derivative (defined in a point) and simultaneously introducing a non-locality in a given space.

It is commonly accepted that the Theory of Thermo-Viscoplasticity (TTV), which plays a central role in the following considerations, began with the publication of Perzyna [22], which, until the present day, serves a basis for many efforts in linking experimental and numerical results for different types of materials. The main results of this theory were discussed in a great number of papers that focused on phenomena such as propagation of mechanical and thermal waves [23,24], viscosity controlled by material parameter [25,26], dispersion [26], or implicit non-locality in the time variable [27]. Nonetheless, the classical TTV formulation does not include directional viscosity, and to include the non-normality extension needs, as all classical plasticity theories, postulation of an additional potential, which is not straightforward and causes the increase of material parameters. Furthermore, the same concern is relevant to the plastic anisotropy effects in terms of the original Perzyna model; to include this effect additional variables and evolution equations for them are needed to be postulated. This limitations were resolved by the generalization of the Perzyna formulation by definition of the fractional flow rule, first proposed in [18] and later developed in [19,28–30].

The implementation of the fractional plastic (rate independent) rule, for the Huber-Mises-Hencky (HMH) yield criterion, in the framework of implicit and explicit procedures and with examples on material point level, was presented in [28]. This was further developed in the subsequent article [29] to any smooth and convex yield criterion but still focusing on rate independent plastic flow. Concluding, in both these articles the non-locality in the stress state was present, however the implicit time non-locality common for the viscoplastic flow was not included in them.

This paper extends the concept of FV, which was first reported in [18], for the Initial Boundary Value Problem (IBVP), and provides a detailed discussion of the model material parameters. The parametric study includes the influence of the overstress power and the relaxation time (which is understood as implicit length scale parameter, as mentioned in [27]) on the dynamic properties of the FV model. Moreover, additional fractional material parameters, which induce the directional viscosity, the non-associative, and the anisotropic plastic flow, were also discussed.

2. Fractional Viscoplasticity

2.1. Remarks on Fractional Calculus

Fractional calculus (FC) introduces a new, universal method for calculating the intensity of changes of various quantities in mathematical models describing experimentally observed phenomena. FC implies a generalization of integer order derivatives, by fractional derivatives (FD). The selection of the FD definition (from an infinite number) can use a type of material as a criterion to obtain the best fitting of the constitutive model to a given experimental evidence. All definitions of the FD have a common property, namely they include summation over an interval abandoning the integer order derivative definition given at a single point; therefore they are called non-local. The classical derivative can be regarded a special case of the FD when its order becomes integer.

In order to explain the FD concept, let us consider a generalized fractional differential operator B_P^α as a composition of fractional integral K_P^α with classical integer (n-th) differential operator [31]

$$B_P^\alpha = K_P^{n-\alpha} \circ \frac{d^n}{dt^n},\tag{1}$$

where α is the order of the derivative, $n = \lfloor \alpha \rfloor + 1$, $\lfloor \cdot \rfloor$ denotes the floor function, P is a parameter set (described below) and $\circ$ denotes the composition operator. B_P^α is referred to as the fractional differential operator B (B-op) of order α and p-set P, and analogously K_P^α identifies the K (K-op) fractional integer operator of order α and p-set P.

The definition of K for the parameter set $P = \langle a, t, b, p, q \rangle$ can be given as

$$\left(K_P^\alpha f\right)(t) = p \int_a^t k_\alpha(t,\tau) f(\tau) d\tau + q \int_t^b k_\alpha(\tau,t) f(\tau) d\tau,\tag{2}$$

where $t \in [a, b]$ and $a < t < b$, p, q are real numbers, and $k_\alpha(t, \tau)$ is a kernel that depends on the order of the derivative α. It can be shown that if k_α is a difference kernel, i.e., $k_\alpha(t, \tau) = k_\alpha(t - \tau)$

and $k_\alpha \in L_1\left([0, b-a]\right)$ then $L_1\left([b, a]\right) \to L_1\left([b, a]\right)$ is well defined, bounded and linear. For explicit definition, the special form of the kernel function can be assumed

$$k_\alpha(t-\tau) = \frac{1}{\Gamma(\alpha)}(t-\tau)^{\alpha-1}, \tag{3}$$

then for $P = \langle a, t, b, 1, 0 \rangle$

$$\left(K_P^\alpha f\right)(t) = \frac{1}{\Gamma(\alpha)}\int_a^t (t-\tau)^{\alpha-1} f(\tau)d\tau = \left({_aI_t^\alpha} f\right)(t), \tag{4}$$

is obtained or, if $P = \langle a, t, b, 0, 1 \rangle$ then

$$\left(K_P^\alpha f\right)(t) = \frac{1}{\Gamma(\alpha)}\int_t^b (\tau-t)^{\alpha-1} f(\tau)d\tau = \left({_tI_b^\alpha} f\right)(t), \tag{5}$$

where Γ is the Euler gamma function. Equations (4) and (5) describe the left and right Riemann-Liouville fractional integrals of the order α, respectively. The application of these operators in Equation (1) leads to the following fractional derivative definitions:

$$\left(B_P^\alpha\right) f(t) = {_a^C D_t^\alpha} f(t) = \frac{1}{\Gamma(n-\alpha)}\int_a^t \frac{f^{(n)}(\tau)}{(t-\tau)^{\alpha-n+1}}d\tau, \tag{6}$$

for $t > a$, and

$$-\left(B_P^\alpha\right) f(t) = {_t^C D_b^\alpha} f(t) = \frac{(-1)^n}{\Gamma(n-\alpha)}\int_t^b \frac{f^{(n)}(\tau)}{(\tau-t)^{\alpha-n+1}}d\tau, \tag{7}$$

for $t < b$. The FD operators ${_a^C D_t^\alpha} f(t)$ and ${_t^C D_b^\alpha} f(t)$ are known as the left- and right-sided Caupto fractional integrals.

Finally, for the purpose of further definition of the FV, the Riesz-Caputo (RC) derivative can be expressed as a linear combination of previously given left and right Caputo derivatives

$$_a^{RC} D_b^\alpha f(t) = \frac{1}{2}\left({_a^C D_t^\alpha} f(t) + (-1)^n {_t^C D_b^\alpha} f(t)\right). \tag{8}$$

It can be shown that for the RC derivative the fundamental property of integer order derivatives is preserved, that is, the derivative of a constant is zero.

2.2. Basic Concepts

In the following section Voigt notation is applied, thus the second rank tensors are ordered as $(6 \times 1$ column matrix)

$$\mathbf{t} = \left(t_{11}\ t_{22}\ t_{33}\ t_{23}\ t_{13}\ t_{12}\right)^{\mathrm{T}} = \left(t_1\ t_2\ t_3\ t_4\ t_5\ t_6\right)^{\mathrm{T}}, \tag{9}$$

whereas the fourth order tensors are represented by 6×6 matrices ordered in accordance with the rule used in Equation (9).

Deformation assumes the additive decomposition of total strain, therefore

$$\varepsilon = \varepsilon^e + \varepsilon^{vp}, \tag{10}$$

or in a rate form

$$\dot{\varepsilon} = \dot{\varepsilon}^e + \dot{\varepsilon}^{vp}, \tag{11}$$

where ε is the total strain, ε^e is the elastic strain and ε^{vp} is the viscoplastic strain. Next, due to thermodynamic restrictions, the elastic strain is related to elastic stress through Hooke's law

$$\sigma^e = \mathcal{L}^e \varepsilon^e, \tag{12}$$

where σ^e denotes the Cauchy stress tensor and $\mathcal{L}^e$ denotes the elastic constitutive tensor. The rate of viscoplastic strain is analogous to the classical viscoplastic definition, namely

$$\dot{\varepsilon}^{vp} = \Lambda \mathbf{p}, \tag{13}$$

where Λ is a scalar multiplier and $\mathbf{p}$ is the second order unit tensor which governs the direction of viscoplastic flow. As the $\mathbf{p}$ tensor is normalized, the magnitude of $\dot{\varepsilon}^{vp}$ depends solely on the Λ parameter.

Following the concept introduced by Perzyna [22], this parameter is expressed as

$$\Lambda = \gamma \left\langle \Phi(F) \right\rangle, \tag{14}$$

where $\gamma = \frac{1}{T_m}$ is the viscosity parameter, Φ is the overstress function that depends on the rate-independent yield surface F, and $\langle \cdot \rangle$ is Macaulay brackets. It is well-known that γ introduces *implicit time non-locality* in the viscoplastic model [27]. Furthermore, the function Φ has the following form

$$\Phi(F) = F^{m_{vp}} = \left(\frac{\sqrt{J_2}}{\kappa} - 1 \right)^{m_{vp}}, \tag{15}$$

where $\sqrt{J_2}$ denotes the second invariant of stress deviator and κ is the static yield stress in simple shear.

Finally, the remaining object needed to be defined is the tensor $\mathbf{p}$. In this place, the difference between the classical theory of viscoplasticity and the new approach is most evident. Let us recall, that in the classical formulation the direction of yield is normal to yield surface and $\mathbf{p}$ can be written as

$$\mathbf{p} = \frac{\partial F}{\partial \sigma} \left(\left\| \frac{\partial F}{\partial \sigma} \right\| \right)^{-1}. \tag{16}$$

It is also well known, that Equations (16) and (15) indicate that the viscoplastic strain is coaxial with the deviatoric stress tensor (associated flow). As a result, the volume change can occur in the range of elastic deformations only. For modern materials such as metal-matrix composites, this assumption is no longer valid. Therefore, the constitutive model should be modified to capture this phenomenon.

The fractional approach assumes the application of the RC operator to $\mathbf{p}$ definition [18]. In such a case, Equation (16) is generalized to the form

$$\mathbf{p} = D^\alpha F \, \|D^\alpha F\|^{-1}, \tag{17}$$

where D^α stands for the RC operator (see Equation (8)). It is worth noting that the proposed formulation of $\mathbf{p}$ introduces the anisotropy of viscoplastic flow and furthermore (due to non-associativity) develops a tool to control the volume change in the plastic range of material behaviour [18]. Another essential remark is that Equation (17) introduces *explicit stress-fractional non-locality* in the overall model. It is important that the thermodynamic restrictions are formulated in a standard manner, and because of complicated structure of Equation (17) they are checked incrementally in the numerical procedure (see [29] for a detailed discussion).

3. Implementation

Introduction of the three-dimensional fractional viscoplastic model requires a numerical procedure that governs the solution. Since our considerations are focused on extreme dynamic processes, the explicit time integration was chosen for finite element method. Therefore, the ABAQUS/Explicit code was utilized together with the user subroutine VUMAT. The critical steps of the implementation are presented below.

In the first step, Hooke's law is written as

$$
\begin{Bmatrix} \sigma_{11} \\ \sigma_{22} \\ \sigma_{33} \\ \sigma_{23} \\ \sigma_{13} \\ \sigma_{12} \end{Bmatrix} = \begin{pmatrix} 2G+\lambda & \lambda & \lambda & 0 & 0 & 0 \\ \lambda & 2G+\lambda & \lambda & 0 & 0 & 0 \\ \lambda & \lambda & 2G+\lambda & 0 & 0 & 0 \\ 0 & 0 & 0 & G & 0 & 0 \\ 0 & 0 & 0 & 0 & G & 0 \\ 0 & 0 & 0 & 0 & 0 & G \end{pmatrix} \begin{Bmatrix} \varepsilon_{11} \\ \varepsilon_{22} \\ \varepsilon_{33} \\ 2\varepsilon_{23} \\ 2\varepsilon_{13} \\ 2\varepsilon_{12} \end{Bmatrix}, \tag{18}
$$

where $G = E/2(1+\nu)$ and $\lambda = E\nu/(1+\nu)(1-2\nu)$ are elastic constants and E and ν denote Young's modulus and Poisson's ratio, respectively. Next, because of application of the HMH yield criterion, the yield function in a matrix form is as follows

$$
F(\sigma) = \begin{Bmatrix} \sigma_{11} \\ \sigma_{22} \\ \sigma_{33} \\ \sigma_{23} \\ \sigma_{13} \\ \sigma_{12} \end{Bmatrix}^T \begin{pmatrix} 1 & -\frac{1}{2} & -\frac{1}{2} & 0 & 0 & 0 \\ -\frac{1}{2} & 1 & -\frac{1}{2} & 0 & 0 & 0 \\ -\frac{1}{2} & \frac{1}{2} & 1 & 0 & 0 & 0 \\ 0 & 0 & 0 & 3 & 0 & 0 \\ 0 & 0 & 0 & 0 & 3 & 0 \\ 0 & 0 & 0 & 0 & 0 & 3 \end{pmatrix} \begin{Bmatrix} \sigma_{11} \\ \sigma_{22} \\ \sigma_{33} \\ \sigma_{23} \\ \sigma_{13} \\ \sigma_{12} \end{Bmatrix} - 3\kappa^2 = 0. \tag{19}
$$

Afterwards, the increment of viscoplastic strain (see Equations (13) and (17)) can be written in the form

$$
\Delta\varepsilon^{vp} = \Delta t \Lambda \mathbf{p} = \Delta t \Lambda \frac{\left\{ \dfrac{D^\alpha F}{\sigma_{11}} \quad \dfrac{D^\alpha F}{\sigma_{22}} \quad \dfrac{D^\alpha F}{\sigma_{33}} \quad \dfrac{D^\alpha F}{\sigma_{23}} \quad \dfrac{D^\alpha F}{\sigma_{13}} \quad \dfrac{D^\alpha F}{\sigma_{12}} \right\}^T}{||D^\alpha F||}, \tag{20}
$$

where (cf. [28])

$$
\begin{Bmatrix} p_{11} \\ p_{22} \\ p_{33} \\ p_{23} \\ p_{13} \\ p_{12} \end{Bmatrix} = \begin{pmatrix} k^M_{(11)} & -\frac{1}{2}k^M_{(11)} & -\frac{1}{2}k^M_{(11)} & 0 & 0 & 0 \\ -\frac{1}{2}k^M_{(22)} & k^M_{(22)} & -\frac{1}{2}k^M_{(22)} & 0 & 0 & 0 \\ -\frac{1}{2}k^M_{(33)} & -\frac{1}{2}k^M_{(33)} & k^M_{(33)} & 0 & 0 & 0 \\ 0 & 0 & 0 & 3k^M_{(23)} & 0 & 0 \\ 0 & 0 & 0 & 0 & 3k^M_{(13)} & 0 \\ 0 & 0 & 0 & 0 & 0 & 3k^M_{(12)} \end{pmatrix} \begin{Bmatrix} \sigma_{11} \\ \sigma_{22} \\ \sigma_{33} \\ \sigma_{23} \\ \sigma_{13} \\ \sigma_{12} \end{Bmatrix} + \begin{Bmatrix} k^Q_{(11)} \\ k^Q_{(22)} \\ k^Q_{(33)} \\ 3k^Q_{(23)} \\ 3k^Q_{(13)} \\ 3k^Q_{(12)} \end{Bmatrix}. \tag{21}
$$

Symbols in Equation (21) denotes

$$
k^M_{(ij)} = \frac{\Gamma(2)}{\Gamma(2-\alpha)} \left[\left(\Delta^L_{(ij)}\right)^{1-\alpha} + \left(\Delta^R_{(ij)}\right)^{1-\alpha} \right], \tag{22}
$$

$$
k^Q_{(ij)} = \left(\frac{\Gamma(2)}{\Gamma(2-\alpha)} - \frac{1}{2}\frac{\Gamma(3)}{\Gamma(3-\alpha)} \right) \left[\left(\Delta^R_{(ij)}\right)^{2-\alpha} - \left(\Delta^L_{(ij)}\right)^{2-\alpha} \right], \tag{23}
$$

$$\Delta^{\mathbf{L}} = \left(\begin{array}{cccccc} \Delta^L_{(11)} & \Delta^L_{(22)} & \Delta^L_{(33)} & \Delta^L_{(23)} & \Delta^L_{(13)} & \Delta^L_{(12)} \end{array} \right)^T, \tag{24}$$

and

$$\Delta^{\mathbf{R}} = \left(\begin{array}{cccccc} \Delta^R_{(11)} & \Delta^R_{(22)} & \Delta^R_{(33)} & \Delta^R_{(23)} & \Delta^R_{(13)} & \Delta^R_{(12)} \end{array} \right)^T, \tag{25}$$

where

$$a_{(ij)} = \sigma_{ij} - \Delta^L_{(ij)}, \quad b_{(ij)} = \sigma_{ij} + \Delta^R_{(ij)}. \tag{26}$$

Terminals $a_{(ij)}, b_{(ij)}$ are needed to define the partial fractional derivatives in Equation (17) that enforce the directional nature of the fractional viscoplastic flow–the subscripts L and R corresponds to the left and the right Caputo derivatives, respectively. In addition, by introducing sections that extend the calculation beyond the material point, a virtual neighbourhood is obtained that results in a non-locality in a stress state.

The interpretation of the virtual surrounding in a stress state depends on the specific material (see [28]), but in general it could be understood as a (homogenized) phenomenological measure of some instability, e.g., for metallic materials it is connected with dislocation nucleation [32–34], nucleation of voids [35] or breakup of grains [36–38] (see review paper [39]). By way of illustration, Figure 1 shows the cross-section of this virtual neighbourhood in the $\sigma_2 - \sigma_3$ plane.

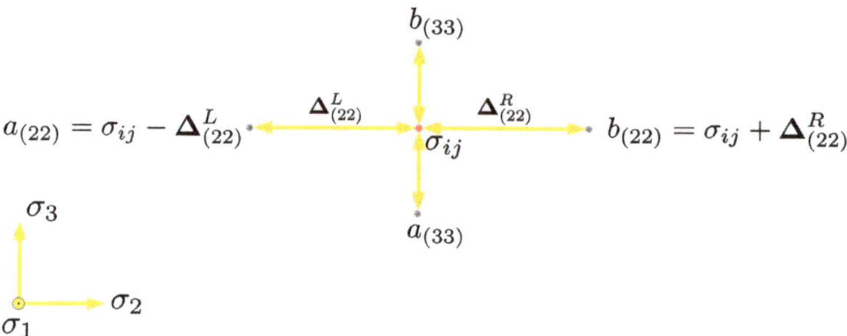

Figure 1. Virtual surrounding of a material point.

The analysis of Equation (21) shows differences in relation to the classical viscoplasticity where the change in volume may only occur in the elastic range—in the classical case, the trace of the **p** tensor equals 0. This condition is abandoned in the fractional formulation (when $\alpha \in (0, 1)$), thus explicitly providing a tool to control the evolution of volume in the plastic range through α and parametric vectors $\Delta^{\mathbf{L}}$ and $\Delta^{\mathbf{R}}$. Moreover, the versatility of the fractional approach is proven for $\alpha = 1$, for which the associated plastic flow as a special case is obtained.

Finally, the flowchart was formulated that presents the general calculation scheme for the elasto-viscoplastic material in the framework of the fractional viscoplastic flow rule for explicit time integration in VUAMT subroutine (see Figure 2). The VUAMT subroutine aims at determination of the values of Cauchy stresses and updating strains and internal variables at time t_{n+1} based on the knowledge of these parameters at the previous moment t_n. The procedure starts with the calculation of the elastic trial stress, which is later used to establish the value of the yield criterion. If this criterion is fulfilled, the plastic multiplier Λ and the direction of plastic flow **p** are calculated according to the flow rule; otherwise the elastic step is conducted.

Given parameters: E, ν, κ, T_m, m

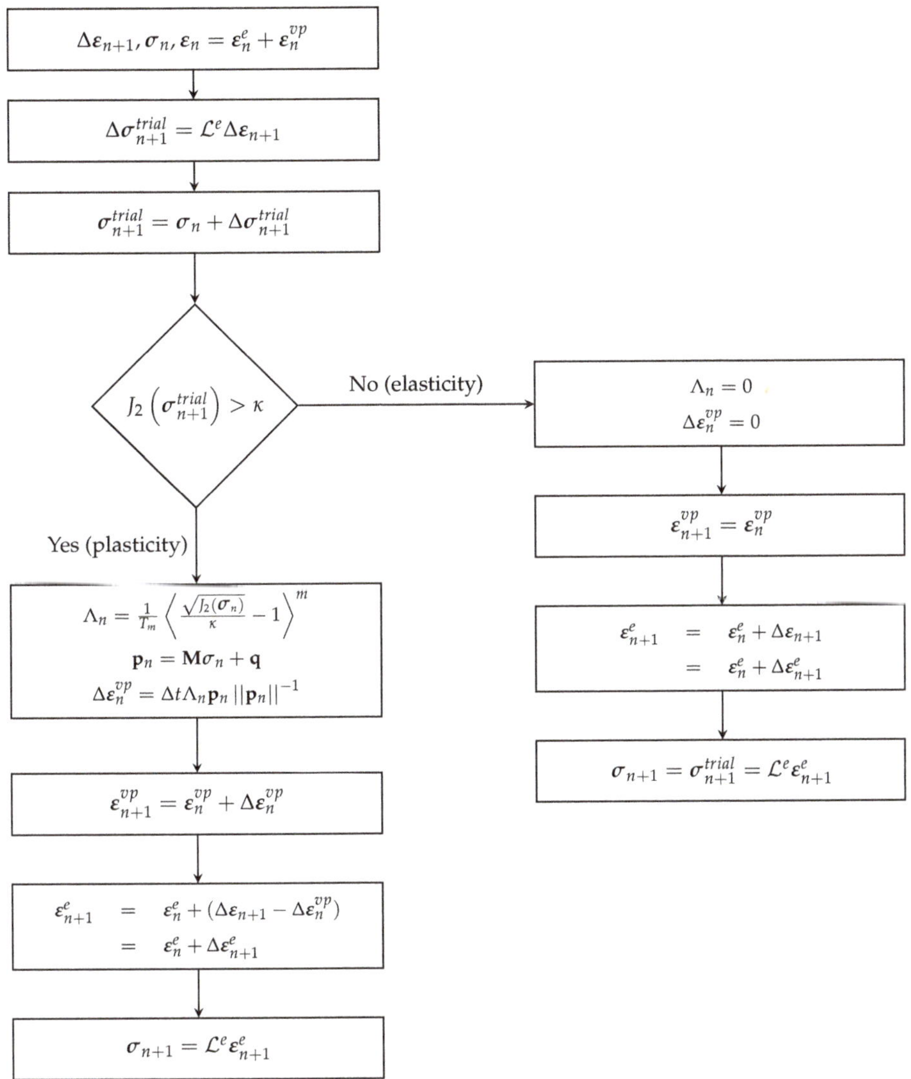

Figure 2. VUMAT subroutine flowchart for the fractional viscoplastic rule.

4. Parametric Study: Uniaxial Tension

4.1. Description of the Numerical Experiment

The conducted parametric study is focused on the material point level represented by a unit cube with dimensions of $1 \times 1 \times 1$ mm discretized by a single finite element C38DR (linear, eight-node brick with reduce integration). The boundary conditions required to achieve uniaxial constraints are shown in Figure 3. Basic mechanical properties were assumed as for the carbon steel, therefore the elastic range was characterized by Young's modulus $E = 205$ GPa and Poisson's ratio $\nu = 0.27$. The fractional flow rule presented in Section 2 was applied in the plastic range. The static yield stress in simple shear for the selected material was $\kappa = 605$ MPa. Other material parameters, depending on the studied case, were chosen as described below.

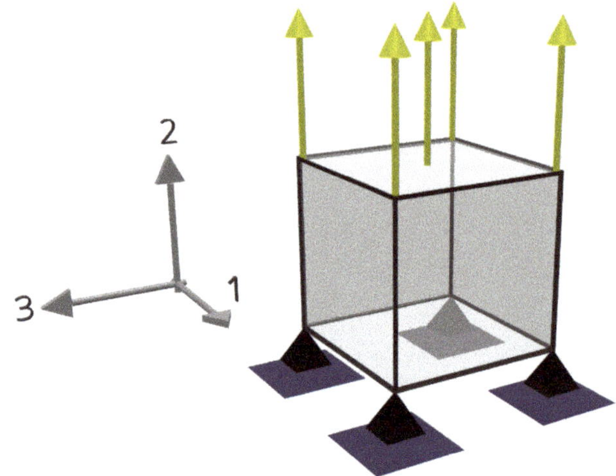

Figure 3. Unit cubic model restricted to uniaxial tension.

The analyzed cases of fractional flow were divided into two groups to show how various combinations of model parameters influence plastic deformation. The first group is focused on the value of the stress-fractional non-locality spread ($\Delta^{L,R}$) and the order (α) of the fractional flow. In the second group, the influence of the material parameters T_m and m under various speeds of the imposed displacement is closely studied. Anticipating the anisotropic behaviour of the fractional material model, these two groups were further subdivided according to the direction where the dominant viscoplastic flow was expected. Hence, two cases were formed for tension direction ($\Delta_{22}^{L,R} = 0.005\kappa \approx 3.0$ MPa) and direction perpendicular to tension ($\Delta_{11}^{L,R} = 0.005\kappa \approx 3.0$ MPa). In each of those cases, other values of the Δ were set to $0.0017\kappa \approx 1.0$ MPa.

Two kinds of plots were used to present the results of the parametric study. The first kind exemplify the relation between three normal strains $\varepsilon_{11}, \varepsilon_{22}, \varepsilon_{33}$. The second type shows the stress–strain relation in the tension (2) direction (it should be pointed out that for this kind of plots, a 'softening' is observed, especially for highest tension velocities, however, this effect is due to the lateral stresses induced by the inertia effects and is not due to constitutive model $\kappa = const$. See Figure 4, where this effect is negligible due to relatively small tension velocity $v = 1\frac{m}{s}$). The research on influence of the fractional derivative order was performed for a set $\alpha \in \{0.1, 0.25, 0.5, 0.75, 0.99, 1.0\}$—as mentioned earlier, fractional generalization of the viscoplasticity reduces to the classical solution for $\alpha = 1$. The study of T_m and m was conducted for three different velocities of tension, i.e., $v = 1, 25$ and $50\ \frac{m}{s}$.

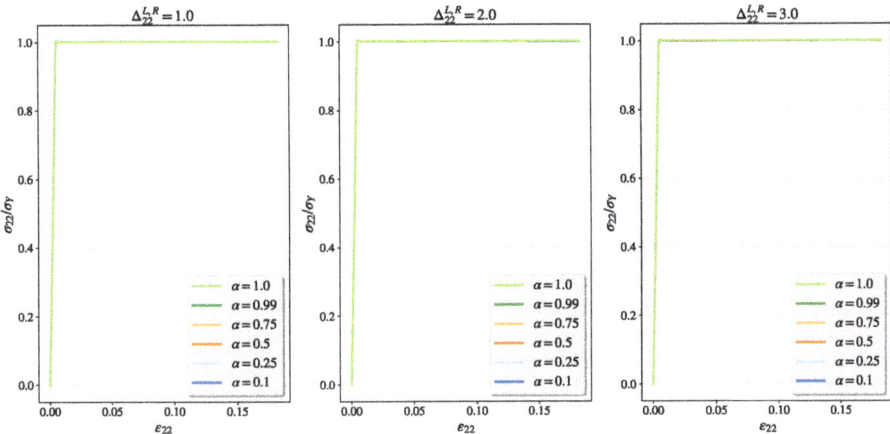

Figure 4. Influence of the order α and the value of material parameter Δ_{22} on the stress–strain relation, for: $v = 1\ \frac{m}{s}, T_m = 2.5e\text{-}6\ s, m = 1$.

4.2. Influence of the Order of FV and Non-Locality in a Stress State on Plastic Flow

4.2.1. Study of Intensified Plastic Flow in Tension Direction for Different Orders of Flow

Figure 4 presents the material response to the applied tension velocity of $v = 1\ \frac{m}{s}$ for different flow intensities in tension direction and flow orders. Increasing the flow intensity parameter $(\Delta^{L,R})$ causes higher evolution of the plastic flow in the chosen direction but in this case the velocity of the load is not sufficient to reveal different behaviour in the stress–strain relation for different values of α (see Figure 4). Next, for the same configuration of material parameters higher velocity is applied, namely $v = 25\ \frac{m}{s}$. At this speed (Figure 5), a slight waveform begins to be visible for $\Delta_{22}^{L,R} = 1.0$ MPa. The amplitude of the stress signal increases with the increases $\Delta_{22}^{L,R}$. Additionally, the influence of α is shown because the decrease in its value translates into greater amplitude of oscillations. So, we conclude that both fractional parameters, control the dynamic properties of a fractional model.

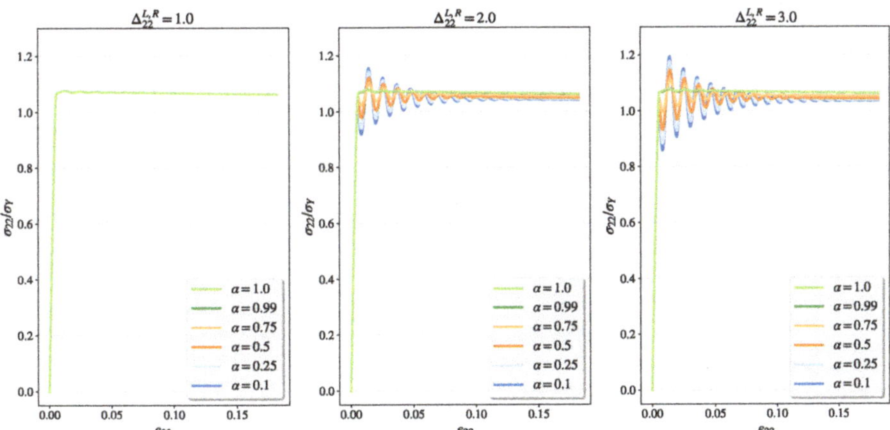

Figure 5. Influence of the order α and the value of material parameter Δ_{22} on the stress–strain relation, for: $v = 25\ \frac{m}{s}, T_m = 2.5e\text{-}6\ s, m = 1$.

4.2.2. Study of Intensified Plastic Flow Perpendicular to the Tension Direction for Different Orders of Flow

Results presented in this section were obtained for a parameter set similar to this in Section 4.2.1 with the difference that flow intensity is increased in the direction perpendicular to tension, namely $\Delta_{11}^{L,R} = 3.0$ MPa. Others components of vectors in Equations (24) and (25) equal 1. As in the discussion in the previous section, the tension velocity of $v = 1\ \frac{m}{s}$ is insufficient to reveal the influence of α on the stress–strain relation (see Figure 6). However, Figure 7 shows that the material prefers to deform in (1) direction when the magnitude of $\Delta_{11}^{L,R}$ growths, hence the $\varepsilon_{11}/\varepsilon_{33}$ ratio is greater then 1. Moreover, the intensity of the flow in the preferred direction increases as the value of α diminishes to 0. Next, as before, the velocity is increased to $v = 25\ \frac{m}{s}$. Figure 8 shows that when the value of $\Delta^{L,R}$ rises, greater amplitude of the oscillation and hardening of the material can be observed. This last effect is inversely proportional to the order of the fractional flow. The relation between $\varepsilon_{11}, \varepsilon_{22}$ and ε_{33} is presented in Figure 9 and is very similar to Figure 7 with the only distinction that slight oscillation occurs as a result of higher velocity. So, we conclude that both fractional parameters control the anisotropic properties of a fractional model in the plastic range.

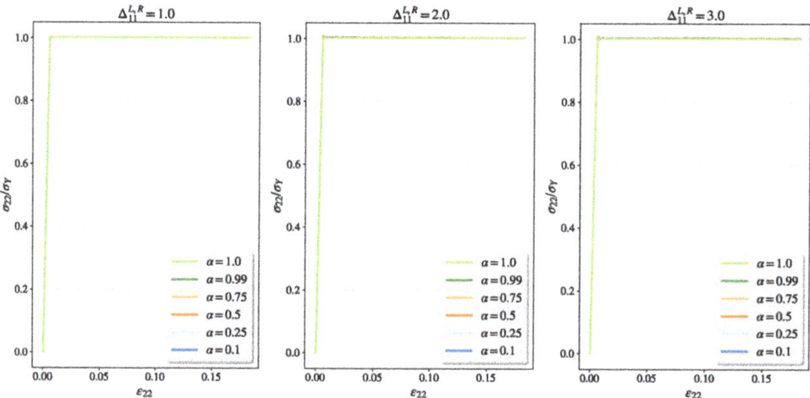

Figure 6. Influence of the order α and the value of material parameter Δ_{11} on the stress–strain relation, for: $v = 1\ \frac{m}{s}, T_m = 2.5e\text{-}6\ s, m = 1$.

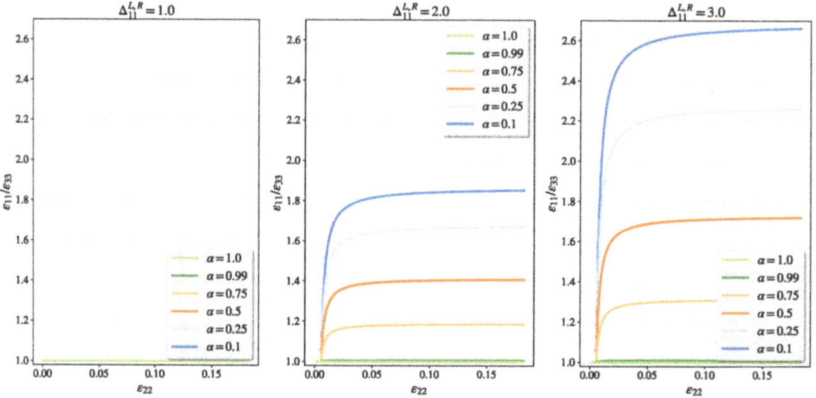

Figure 7. Influence of the order α and the value of material parameter Δ_{11} on the relation between three normal stresses, for: $v = 1\ \frac{m}{s}, T_m = 2.5e\text{-}6\ s, m = 1$.

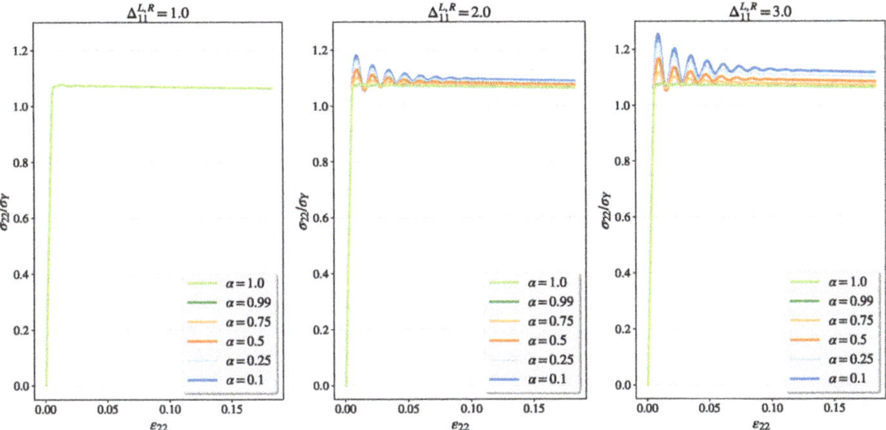

Figure 8. Influence of the order α and the value of material parameter Δ_{11} on the stress–strain relation, for: $v = 25\ \frac{m}{s}, T_m = 2.5e\text{-}6\ s, m = 1$.

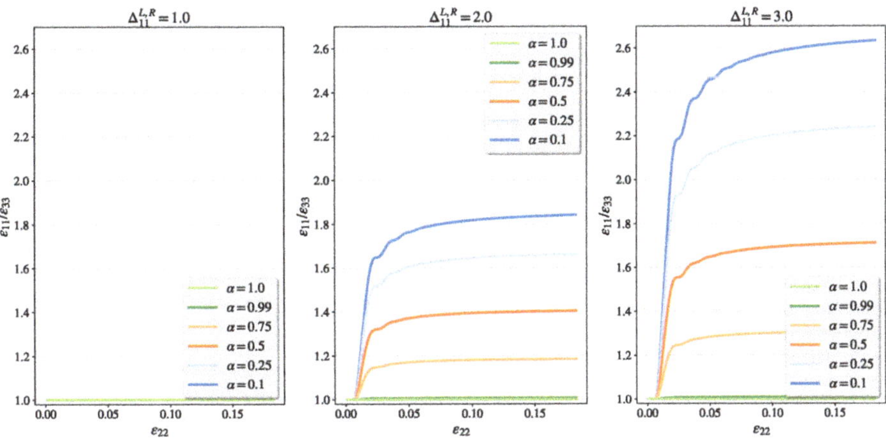

Figure 9. Influence of the order α and the value of material parameter Δ_{11} on the relation between three normal stresses, for: $v = 25\ \frac{m}{s}, T_m = 2.5e\text{-}6\ s, m = 1$.

4.3. Influence of the Relaxation Time and the Overstress Power

4.3.1. Study of the Fractional Flow Under Different Dynamic Loading Rates for Intensified Plastic Flow in Tension Direction

Here we assume that the intensified plastic flow, determined by $\Delta_{22}^{L,R} = 3.0$ MPa, is in tension direction. Figure 10 presents the effect of different relaxation times for various velocities of tension. It should be pointed out that in order to increase clarity of interpretation both the classical ($\alpha = 1$) and the fractional ($\alpha = 0.75$) solutions are compared on each graph. As can be seen, when the relaxation time grows, the hardening of the material as well as the stress waves oscillations increase. The latter is especially pronounced for the relaxation time $T_m = 2.5e\text{-}5\ s$.

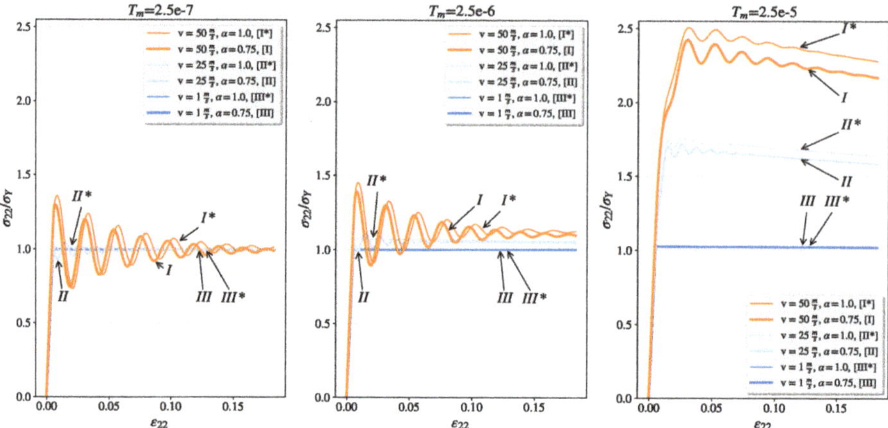

Figure 10. Influence of the relaxation parameter T_m and the value of applied velocity field v on the stress–strain relation, for: $\alpha = 0.75, m = 1, \Delta_{22} = 3.0$.

Figure 11 presents the result of increasing the value of the overstress parameter m. For the fractional ($\alpha = 0.75$) viscoplastic material the stress level is smaller in relation to the classical ($\alpha = 1.0$) viscoplastic solution (Figures 10 and 11).

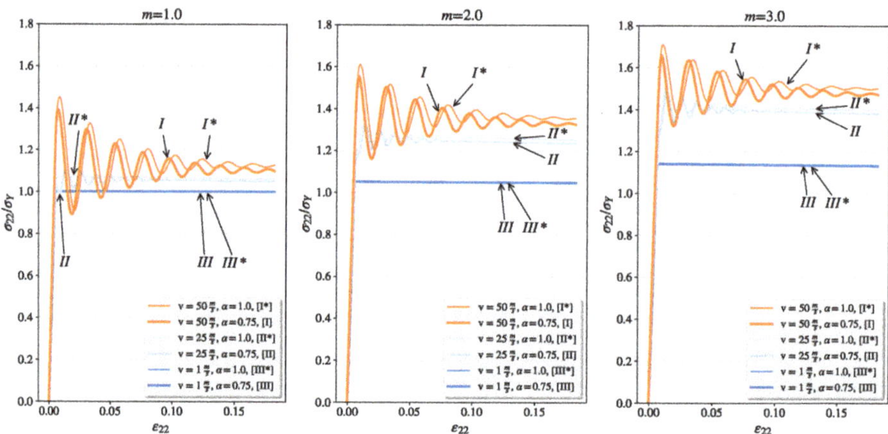

Figure 11. Influence of the material parameter m and the value of applied velocity field v on the stress–strain relation, for: $\alpha = 0.75, T_m = 2.5e\text{-}6$ s, $\Delta_{22} = 3.0$.

Results discussed above indicate that the relaxation time and overstress power, together with fractional parameters, control the level of strain rate hardening and stress waves oscillation amplitude.

4.3.2. Study of the Fractional Flow Under Different Dynamic Loading for the Intensified Plastic Flow Perpendicular to the Tension Direction

In this section, it is assumed that fractional flow is intensified in the direction perpendicular to tension load ($\Delta_{11}^{L,R} = 3.0\ MPa$). Figure 12 shows that raising relaxation time causes similar effects to those discussed in the previous section. The biggest change occurs for the relaxation time $T_m = 2.5e\text{-}5$ s both in the stress level and the stress wave oscillations frequency. In Figure 13 a clear anisotropy of the plastic deformation for $\alpha = 0.75$ can be noticed. As before, it is observed that increasing the value of m causes a growth in the material strain hardening without any apparent influence on the frequency of the stress wave (see Figure 14). By analogy to what is presented for T_m, in Figure 15 the dominant nature of ε_{11} can be observed, in addition to more pronounced oscillations for greater values of m. The analysis of the $\sigma - \varepsilon$ relation for $\alpha = 1$ and $\alpha = 0.75$ also revealed that the stress levels of the fractional model are generally greater than for the classical solution (see Figures 12 and 14). This last observation is different from what was discovered in the previous section.

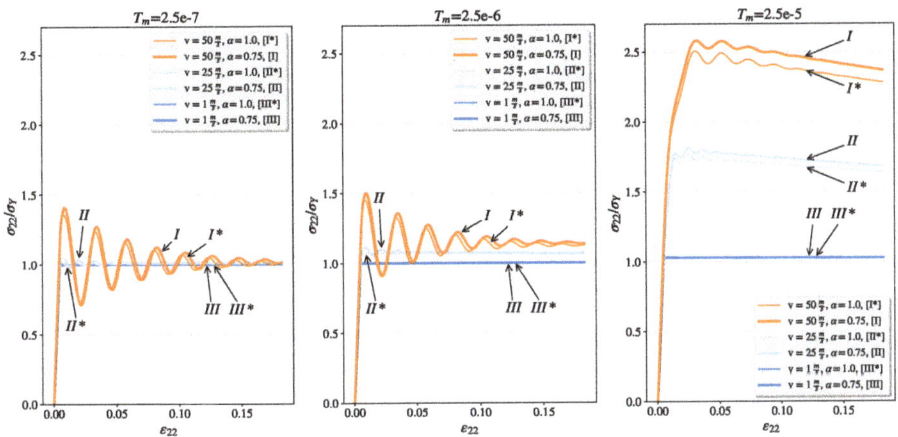

Figure 12. Influence of the relaxation parameter T_m and the value of applied velocity field v on the stress–strain relation, for: $\alpha = 0.75, m = 1, \Delta_{11} = 3.0$.

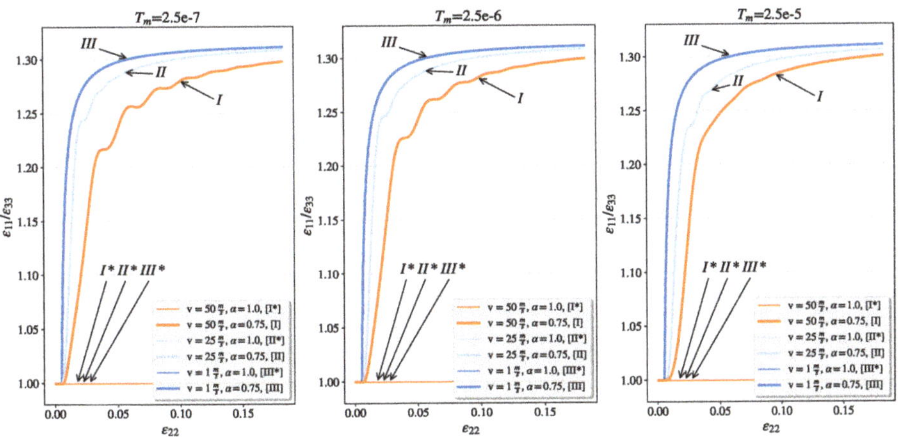

Figure 13. Influence of the relaxation parameter T_m and the value of applied velocity field v on the relation between three normal stresses, for: $\alpha = 0.75, m = 1, \Delta_{11} = 3.0$.

In conclusion, as before, a prominent influence of the relaxation time and overstress power, together with fractional parameters, on the level of strain rate hardening and stress waves oscillation amplitude is observed. Additionally, the impact on dynamic properties of deformation anisotropy for fractional material was confirmed.

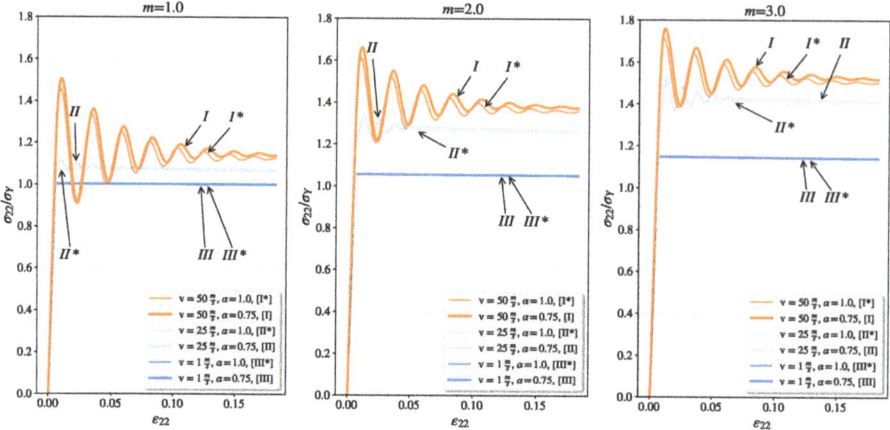

Figure 14. Influence of the material parameter m and the value of applied velocity field v on the stress–strain relation, for: $\alpha = 0.75, T_m = 2.5e\text{-}6\ s, \Delta_{11} = 3.0$.

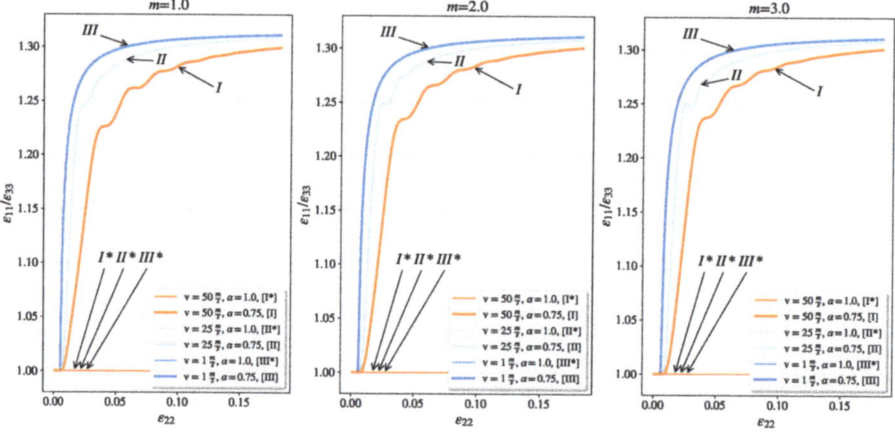

Figure 15. Influence of the material parameter m and the value of applied velocity field v on the relation between three normal stresses, for: $\alpha = 0.75, T_m = 2.5e\text{-}6\ s, \Delta_{11} = 3.0$.

4.4. Study of the Disperse Character of the Fractional Viscoplastic Stress Waves

The last study performed was the analysis of the disperse character of the fractional viscoplastic stress waves. As shown in the previous sections, the uniaxial dynamic deformation induces a stress waves. The regularity of the oscillations was determined by averaging intervals between peaks and then calculating the frequency. Table 1 lists the stress wave frequencies for $\Delta_{22}^{L,R} = 3.0$ MPa and $\Delta_{11}^{L,R} = 3.0$ MPa, which were depicted in the middle graphs of Figures 10 and 12. The material parameter $\Delta^{L,R}$ does not appear in the classical viscoplasticity, so there is no change in the stress wave frequency for various values of this parameter when $\alpha = 1$, thus tension velocity is only important. However, for fractional material, i.e., when $\alpha = 0.75$, both flow intensity parameters and tension velocities modulate the frequency of stress waves. For $\alpha = 0.75$ the frequencies are higher when the distinguished direction is co-linear with tension ($\Delta_{22} = 3.0$ MPa).

Table 1. The stress wave frequencies for $T_m = 2.5e\text{-}6\ s$, $m = 1$

		$\Delta_{22} = 3.0$	$\Delta_{11} = 3.0$
$v = 25\ \frac{m}{s}$	$\alpha = 1$	2.085 MHz	2.085 MHz
	$\alpha = 0.75$	2.108 MHz	1.996 MHz
$v = 50\ \frac{m}{s}$	$\alpha = 1$	2.073 MHz	2.073 MHz
	$\alpha = 0.75$	2.157 MHz	2.028 MHz

The results presented in Table 2 correspond to the investigation of the role of the relaxation time and overstress power discussed in Section 4.3. Regardless of the value of α, the largest change in frequency can be observed between columns 2 and 3, that is for $T_m = 2.5e\text{-}6\ s$ and $T_m = 2.5e\text{-}5\ s$. The comparison of the values in columns 1 ($T_m = 2.5e\text{-}7\ s$) and 2 ($T_m = 2.5e\text{-}6\ s$) shows that the stress wave frequencies are the same or very similar. The impact of m shows that the significant influence of this parameter was only recorded for $T_m = 2.5e\text{-}5\ s$. For both $\alpha = 0.75$ and $\alpha = 1$ the increase of the overstress power results in the increase of the frequency of the stress wave.

Table 2. The stress wave frequencies for $\Delta_{22} = 3.0\ MPa$, $v = 50\ \frac{m}{s}$

		T_m		
		2.5e-7	2.5e-6	2.5e-5
$m = 1$	$\alpha = 1$	2.073 MHz	2.073 MHz	2.274 MHz
	$\alpha = 0.75$	2.157 MHz	2.157 MHz	2.288 MHz
$m = 2$	$\alpha = 1$	2.073 MHz	2.085 MHz	2.182 MHz
	$\alpha = 0.75$	2.157 MHz	2.157 MHz	2.207 MHz
$m = 3$	$\alpha = 1$	2.073 MHz	2.085 MHz	2.169 MHz
	$\alpha = 0.75$	2.157 MHz	2.157 MHz	2.182 MHz

One concludes that the relaxation time and overstress power, together with fractional parameters, control the dispersive character of stress waves, and even more, makes this attribute directional.

5. Conclusions

The analysis of the dynamic properties of the Perzyna model of viscoplasticity (implicit time, non-local) generalized using fractional calculus (explicit, stress-fractional, non-local) leads to the following conclusions:

- Fractional viscoplasticity introduces an additional set of material parameters, namely flow order α and virtual stress state surrounding Δ.
- Fractional parameters α and Δ control the dynamic properties of the fractional model, especially hardening, the character of the stress waves, and plastic anisotropy.
- The direction of the flow vector is controlled by Δ, which in general leads to non-normality of plastic flow.
- As in the classical Perzyna model, the relaxation time T_m and the overstress power m affect the strain rate hardening and the character of the stress waves.

- Induced plastic anisotropy of the fractional model should be regarded not only in the classical sense as directional deformation but also as directional viscosity, which results in directional dispersive character.

The above results are fundamental from the point of view of modeling strain localization and damage phenomena. Both these aspects will serve as a base for future studies.

Author Contributions: All of the authors have contributed to the writing of this paper. They read and approved the manuscript.

Funding: The National Science Centre, Poland under Grant No. 2017/27/B/ST8/00351.

Acknowledgments: This work is supported by the National Science Centre, Poland under Grant No. 2017/27/B/ST8/00351.

Conflicts of Interest: The authors declare no conflict of interest.

References

1. Podlubny, I. Fractional Differential Equations. In *Mathematics in Science and Engineering*; Academin Press: Cambridge, MA, USA, 1999; Volume 198.
2. Samko, S.; Kilbas, A.; Marichev, O. *Fractional Integrals and Derivatives: Theory and Applications*; Gordon and Breach: Amsterdam, The Netherlands, 1993.
3. Meng, R.; Yin, D.; Zhou, C.; Wu, H. Fractional description of time-dependent mechanical property evolution in materials with strain softening behavior. *Appl. Math. Modell.* **2016**, *40*, 398–406. [CrossRef]
4. Li, M. Three Classes of Fractional Oscillators. *Symmetry* **2018**, *10*, 40. [CrossRef]
5. Zhao, J.; Zheng, L.; Zhang, X.; Liu, F. Convection heat and mass transfer of fractional MHD Maxwell fluid in a porous medium with Soret and Dufour effects. *Int. J. Heat Mass Trans.* **2016**, *103*, 203–210. [CrossRef]
6. Jinhu, Z.; Liancun, Z.; Xinxin, Z.; Fawang, L. Mixed convection heat transfer of viscoelastic fluid along an inclined plate obeying the fractional constitutive laws. *Heat Trans. Res.* **2017**, *48*, 1165–1178.
7. Zhang, X.; Liu, L.; Wu, Y. The uniqueness of positive solution for a fractional order model of turbulent flow in a porous medium. *Appl. Math. Lett.* **2014**, *37*, 26–33. [CrossRef]
8. Cherniha, R.; Gozak, K.; Waniewski, J. Exact and Numerical Solutions of a Spatially-Distributed Mathematical Model for Fluid and Solute Transport in Peritoneal Dialysis. *Symmetry* **2016**, *8*, 50. [CrossRef]
9. Atanackovic, T.; Janev, M.; Oparnica, L.; Pilipovic, S.; Zorica, D. Space–time fractional Zener wave equation. *Proc. Math. Phys. Eng. Sci.* 2015, 471, 20140614.
10. Ren, T.; Li, S.; Zhang, X.; Liu, L. Maximum and minimum solutions for a nonlocal p-Laplacian fractional differential system from eco-economical processes. *Bound. Value Probl.* **2017**, *2017*, 118. [CrossRef]
11. Sumelka, W.; Voyiadjis, G. A hyperelastic fractional damage material model with memory. *Int. J. Solids Struct.* **2017**, *124*, 151–160. [CrossRef]
12. Klimek, M. Fractional sequential mechanics—Models with symmetric fractional derivative. *Czechoslov. J. Phys.* **2001**, *51*, 1348–1354. [CrossRef]
13. Drapaca, C.; Sivaloganathan, S. A Fractional Model of Continuum Mechanics. *J. Elast.* **2012**, *107*, 107–123. [CrossRef]
14. Sumelka, W.; Szajek, K.; Łodygowski, T. Plane strain and plane stress elasticity under fractional continuum mechanics. *Arch. Appl. Mech.* **2015**, *89*, 1527–1544. [CrossRef]
15. Tomasz, B. Analytical and numerical solution of the fractional Euler–Bernoulli beam equation. *J. Mech. Mater. Struct.* **2017**, *12*, 23–34.
16. Lazopoulos, K.; Lazopoulos, A. Fractional vector calculus and fluid mechanics. *J. Mech. Behav. Mater.* **2017**, *26*, 43–54. [CrossRef]
17. Peter, B. Dynamical Systems Approach of Internal Length in Fractional Calculus. *Eng. Trans.* **2017**, *65*, 209–215.
18. Sumelka, W. Fractional viscoplasticity. *Mech. Res. Commun.* **2014**, *56*, 31–36. [CrossRef]
19. Sun, Y.; Shen, Y. Constitutive model of granular soils using fractional-order plastic-flow rule. *Int. J. Geomech.* **2017**, *17*, 04017025. [CrossRef]
20. Sun, Y.; Xiao, Y. Fractional order model for granular soils under drained cyclic loading. *Int. J. Numer. Anal. Meth. Geomech.* **2017**, *41*, 555–577. [CrossRef]

21. Sun, Y.; Xiao, Y. Fractional order plasticity model for granular soils subjected to monotonic triaxial compression. *Int. J. Solids Struct.* **2017**, *118-119*, 224–234. [CrossRef]

22. Perzyna, P. The constitutive equations for rate sensitive plastic materials. *Q. Appl. Math.* **1963**, *20*, 321–332. [CrossRef]

23. Glema, A.; Łodygowski, T. On importance of imperfections in plastic strain localization problems in materials under impact loading. *Arch. Mech.* **2002**, *54*, 411–423.

24. Glema, A. Analysis of wave nature in plastic strain localization in solids. In *Rozprawy*; Publishing House of Poznan University of Technology: Poznan, Poland, 2004; Volume 379. (In Polish)

25. Glema, A.; Łodygowski, T.; Perzyna, P. Interaction of deformation waves and localization phenomena in inelastic solids. *Comput. Meth. Appl. Mech. Eng.* **2000**, *183*, 123–140. [CrossRef]

26. Glema, A.; Łodygowski, T.; Perzyna, P. Localization of plastic deformations as a result of wave interaction. *Comput. Assist. Mech. Eng. Sci.* **2003**, *10*, 81–91.

27. Perzyna, P. The Thermodynamical Theory of Elasto-Viscoplasticity. *Eng. Trans.* **2005**, *53*, 235–316.

28. Sumelka, W.; Nowak, M. Non-normality and induced plastic anisotropy under fractional plastic flow rule: A numerical study. *Int. J. Numer. Anal. Meth. Geomech.* **2016**, *40*, 651–675. [CrossRef]

29. Sumelka, W.; Nowak, M. On a general numerical scheme for the fractional plastic flow rule. *Mech. Mater.* **2017**, *116*, 120–129. [CrossRef]

30. Xiao, R.; Sun, H.; Chen, W. A finite deformation fractional viscoplastic model for the glass transition behavior of amorphous polymers. *Int. J. Non-Linear Mech.* **2017**, *93*, 7–14. [CrossRef]

31. Odzijewicz, T.; Malinowska, A.; Torres, D. Green's theorem for generalized fractional derivatives. *Fract. Calc. Appl. Anal.* **2013**, *16*, 64–75. [CrossRef]

32. Ziegler, H. *An Introduction to Thermomechanics*; North Holland Series in Applied Mathematics and Mechanics; North-Holland Publishing Company: North Holland, The Netherlands, 1983; Volume 21.

33. Dao, M.; Asaro, R. Non-Schmid effects and localized plastic flow in intermetallic alloys. *Mater. Sci. Eng. A* **1993**, *170*, 143–160. [CrossRef]

34. Racherla, V.; Bassani, J. Strain burst phenomena in the necking of a sheet that deforms by non-associated plastic flow. *Modell. Simul. Mater. Sci. Eng.* **2007**, *15*, S297–S311. [CrossRef]

35. Marin, E.; McDowell, D. Models for Compressible Elasto-Plasticity Based on Internal State Variables. *Int. J. Damage Mech.* **1998**, *7*, 47–83. [CrossRef]

36. Leffers, T. Lattice rotations during plastic deformation with grain subdivision. *Mater. Sci. Forum* **1994**, *157–162*, 1815–1820. [CrossRef]

37. Hughes, D.; Liu, Q.; Chrzan, D.; Hansen, N. Scaling of microstructural parameters: Misorientations of deformation induced boundaries. *Acta Mater.* **1997**, *45*, 105–112. [CrossRef]

38. Steinmann, P.; Kuhl, E.; Stein, E. Aspects of non-associated single crystal plasticity: Influence of non-schmid effects and localization analysis. *Int. J. Solids Struct.* **1998**, *35*, 4437–4456. [CrossRef]

39. McDowell, D. Viscoplasticity of heterogeneous metallic materials. *Mater. Sci. Eng. R* **2008**, *62*, 67–123. [CrossRef]

MDPI

St. Alban-Anlage 66

4052 Basel

Switzerland

Tel. +41 61 683 77 34

Fax +41 61 302 89 18

www.mdpi.com

Symmetry Editorial Office

E-mail: symmetry@mdpi.com

www.mdpi.com/journal/symmetry